Springer Series in Advanced Manufacturing

Series Editor

Duc Truong Pham, University of Birmingham, Birmingham, UK

The **Springer Series in Advanced Manufacturing** includes advanced textbooks, research monographs, edited works and conference proceedings covering all major subjects in the field of advanced manufacturing.

The following is a non-exclusive list of subjects relevant to the series:

1. Manufacturing processes and operations (material processing; assembly; test and inspection; packaging and shipping).
2. Manufacturing product and process design (product design; product data management; product development; manufacturing system planning).
3. Enterprise management (product life cycle management; production planning and control; quality management).

Emphasis will be placed on novel material of topical interest (for example, books on nanomanufacturing) as well as new treatments of more traditional areas.

As advanced manufacturing usually involves extensive use of information and communication technology (ICT), books dealing with advanced ICT tools for advanced manufacturing are also of interest to the Series.

Springer and Professor Pham welcome book ideas from authors. Potential authors who wish to submit a book proposal should contact Anthony Doyle, Executive Editor, Springer, e-mail: anthony.doyle@springer.com.

More information about this series at http://www.springer.com/series/7113

Apoorva Shastri · Aniket Nargundkar ·
Anand J. Kulkarni

Socio-Inspired Optimization Methods for Advanced Manufacturing Processes

 Springer

Apoorva Shastri
Symbiosis Institute of Technology
Symbiosis International (Deemed
University)
Pune, India

Aniket Nargundkar
Symbiosis Institute of Technology
Symbiosis International (Deemed
University)
Pune, India

Anand J. Kulkarni
Symbiosis Institute of Technology
Symbiosis International (Deemed
University)
Pune, India

ISSN 1860-5168   ISSN 2196-1735   (electronic)
Springer Series in Advanced Manufacturing
ISBN 978-981-15-7799-4   ISBN 978-981-15-7797-0   (eBook)
https://doi.org/10.1007/978-981-15-7797-0

This Springer imprint is published by the registered company Springer Nature Singapore Pte Ltd.
The registered company address is: 152 Beach Road, #21-01/04 Gateway East, Singapore 189721,
Singapore

Authors would like to dedicate this book to their family members for their valuable support in all respect.

Preface

The book focuses on over five years of extensive theoretical, simulation and experimental work on two niche aspects of manufacturing and associated solution methodology. The first aspect is associated with advanced manufacturing processes such as electric discharge machining and micro-machining methods such as micro-turning, micro-milling and micro-drilling, abrasive water jet machining as well as machining of CFRP composites for aerospace applications. The second aspect is associated with the solution to the above problems using several variants of emerging AI-based socio-inspired method referred to as cohort intelligence. The book exhaustively discusses these advanced manufacturing processes including illustrative presentations of the processes, mathematical modeling, the need to optimize associated parameter problems, etc. In addition, the cohort intelligence methodology and its several variants including multi-cohort intelligence have been thoroughly discussed along with illustrative examples. The solutions have been presented using over 50 experimentally achieved plots and figures and illustrations along with over 25 tables. This may help in providing a complete understanding of the framework. The theoretical and statistical rigor is validated by comparing the solutions against evolutionary algorithm such as genetic algorithm as well as simulation annealing, response surface methodology, firefly algorithm, particle swarm optimization and experimental work. In addition, the proposed book also discusses a critical review of numerous socio-inspired optimization methods.

The mathematical level in all the chapters is well within the grasp of the scientists as well as the undergraduate and graduate students from the mechanical engineering and computer science domains. The reader is encouraged to have basic knowledge of mathematical analysis, mechanical engineering and manufacturing processes. Every problem as well as the technique presented in the book has been coded in MATLAB software. All the executable codes are provided on sites www.google.com/site/oatresearch/home.

The authors would like to thank Dr. Ramesh Nath Premnath, Editor Applied Sciences; Dr. Pham Duc Truong, Series Editor, Springer Series in Advanced Manufacturing; Springer Nature Singapore for the editorial assistance and excellent cooperative collaboration to produce this important scientific work. We hope that the reader will share our excitement to present this volume on applying several variations of cohort intelligence algorithm for solving practically important advanced manufacturing processes and will find it useful.

Pune, India

Apoorva Shastri
Aniket Nargundkar
Anand J. Kulkarni

Contents

Chapter 1
Introduction to Advanced Manufacturing Processes and Optimization Methodologies

Manufacturing can be defined as the application of mechanical, physical, and chemical processes to convert the geometry, properties, and/or shape of raw material into finished parts or products. This includes all intermediate processes required for the production and integration of the final product. Manufacturing involves interrelated activities which include product design, material selection, production process planning, production, quality assurance, management and marketing of products. The contemporary manufacturing processes have become extremely complex owing to the technological advances in last few decades. The materials and processes initially used to manufacture the products by casting and smithy have been gradually developed over the centuries, using new materials and more complex operations at the increasing rates of production and enriched quality [29].

Rao [29] has divided the manufacturing processes into five categories such as primary shaping processes, machining processes, forming processes, joining processes, and surface finish processes.

1. **Primary Process**: This process is used to give the raw material a pre-defined shape which involves casting process, plastic forming technology, powder metallurgy, bending of the metals, forging, gas cutting, etc.
2. **Machining process**: Primary process may not results into the complete finished goods. Machining is one of the processes of making the primary shapes into finished product. The process to remove unwanted materials from the material using cutting tools is known as machining. The purpose is to give perfect dimension and desired shape.
3. **Metal Forming process**: Metal forming is another manufacturing technique used to give the metal desired shape through the application of pressure/stress, forces like compression or tension or shear or the combination of both. By applying

A. Shastri et al., *Socio-Inspired Optimization Methods for Advanced Manufacturing Processes*, Springer Series in Advanced Manufacturing, https://doi.org/10.1007/978-981-15-7797-0_1

such forces permanent deformation of the metal is achieved. These processes are also known as mechanical working processes.

4. **Joining Processes**: The joining processes are needed to make air-tight joints and semi-permanent as well as permanent fasteners. These joining techniques are vital in making large assemblies.

5. **Surface Finishing**: It is a process which is applied to get the desired levelling of the surface and also to obtain the desired smoothness. Very negligible amount of the materials is removed from the surface by this process and does not involve any effective change in the dimension and shape.

1.1   Advanced Manufacturing Processes

The primary cutting mechanism or material removal mechanism of conventional manufacturing processes such as turning, milling, drilling, grinding, etc. is based on the material removal by chip formation process. The 20th century has seen the introduction of new generation of materials and alloys comprising of special characteristics such as high strength-to-weight ratio, high stiffness and toughness, high heat capacity and thermal conductivity, etc. due to the demand from various applications. The machining of advanced materials has brought new challenges such as rapid deterioration of the cutting tools, inferior quality of machined parts, etc. The major reasons behind these problems are generation of high temperatures and stresses during machining of these advanced materials [32]. This has led to the development and establishment of the Non Traditional Machining (NTM) or Advanced Manufacturing Processes (AMP) as efficient and economic alternatives to the conventional ones. Today, NTMs/AMPs with varied machining capabilities and specifications are available for a wide range of applications. Abrasive Water Jet Machining (AWJM), Electric Discharge Machining (EDM), Laser Beam Machining (LBM), Ultrasonic Machining (USM), Electro Chemical Machining (ECM) are some of the examples of the AMPs. Moreover, mechanical micro machining such as micro turning, micro milling, and micro drilling classify as AMPs. Some of the researchers also classify sustainable manufacturing such as Minimum Quantity Lubrication (MQL), and Cryogenic Machining as AMPs.

AMPs have certain specific characteristics which are distinct from the conventional processes. Material removal in AMPs may occur with or without chip formation such as in AWJM, chips are of microscopic size and in case of ECM material removal occurs due to electrochemical dissolution at atomic level. Another distinct feature is that in AMPs, a physical tool may not be present such as in LBM, machining is carried out by laser beam. Most of the AMPs do not necessarily utilize mechanical energy for material removal mechanism, however uses a different energy domains for machining. For example, in USM, and AWJM, mechanical energy is used to machine material, whereas in ECM electrochemical dissolution constitutes material removal [20, 29]. Thus, generally classification of NTM processes is carried out depending on the nature of energy used for material removal. Figure 1.1 describes

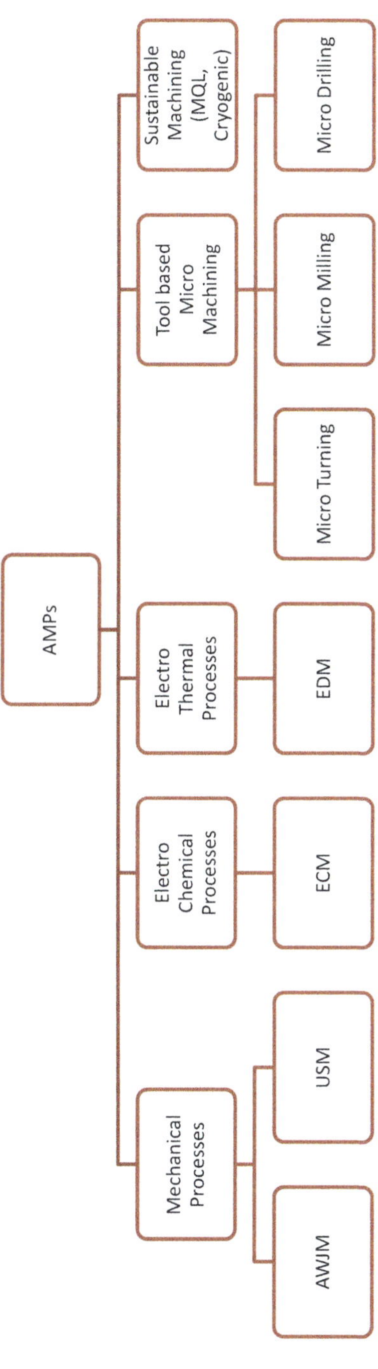

Fig. 1.1 Classification of AMPs

the classification of various AMPs based on the mechanism of material removal and energy used.

This book focuses on the optimization of the AMPs such as AWJM, EDM, Micro Machining processes viz. micro turning, micro drilling, and micro milling. Moreover, the real world application of micro drilling of Carbon Fibre Reinforcement Plastic (CFRP) composite for aerospace application is also illustrated. The remainder of the chapter is organized as follows. Section 1.2 discusses the need for modelling and optimization of AMPs. Section 1.3 describes the EDM process. AWJM process has been discussed in Sect. 1.4. Tool based Micro Machining processes have been described in Sect. 1.5. A detailed comprehensive literature review has also been presented at the end of each section.

1.2   Need for Modelling and Optimization of Advanced Manufacturing Processes

As discussed earlier, the manufacturing processes have become extremely complex owing to the recent technological advances. Today's manufacturing industry is trying to address the challenges such as growing needs for safety, reduced time-to-market that implies short manufacturing time, minimal manufacturing costs through the efficient use of the resources and, expected quality of highly customized products. Determining optimal process parameter settings critically influences productivity, quality, and cost of production. Therefore, optimal manufacturing process parameter setting is recognized as one of the most important activity. Earlier, engineers practiced trial-and-error approach which depends on the experience and intuition to determine initial process parameter settings. However, this approach is not suitable in the context of today's manufacturing processes. Consequently, the optimization of manufacturing process parameters requires new reliable methods and approaches, based on modelling activities, in order to predict the output of the manufacturing processes or in other words to predict the behaviour of the manufacturing process [9].

1.3   Electrical Discharge Machining (EDM)

EDM is an electro-thermal, Advanced Manufacturing Process in which electrical energy is used to generate spark between tool and workpiece and material is removed. According to [20], EDM is mainly used to machine high strength temperature resistant materials and alloys with intricate geometries. The material removal in EDM is based on the erosion of unwanted material from the workpiece. In EDM, a potential difference is applied between the tool (-ve terminal) and work piece (+ve terminal) immersed in a dielectric medium such as kerosene or deionized water, as illustrated

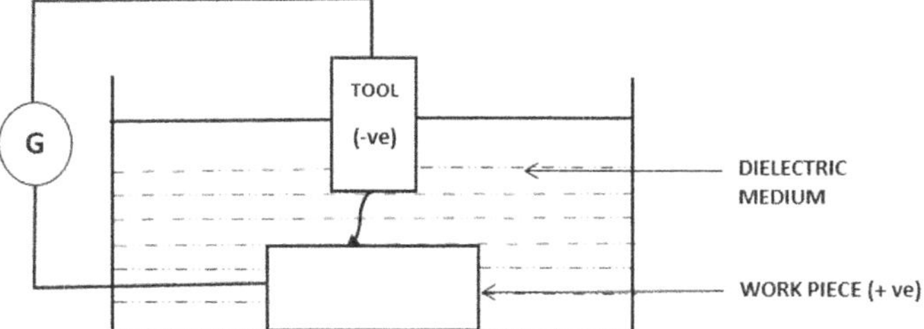

Fig. 1.2 Illustration of EDM process

in Fig. 1.2. Depending upon the applied potential difference and the gap between the tool and work piece, an electric field is established. As the electric field is established between the tool and the work piece, the free electrons on the tool are subjected to electrostatic forces and gain a high velocity and energy. These electrons collide with the dielectric molecules which results into ionization. Thus, as the electrons get accelerated, more positive ions and electrons would get generated. This cyclic process would increase the concentration of electrons and ions in the dielectric medium between the tool and the work piece at the spark gap. Such movement of electrons and ions can be visually seen as a spark. Thus the electrical energy is dissipated as the thermal energy of the spark. The high speed electrons then impinge on the work piece and ions on the tool. According to the experimental investigations by Muthuramalingam and Mohan [25] and Gopalakannan and Senthilvelan [14], during this process electrode wear takes place and hence, electrode wear rate becomes critical to control.

Table 1.1 below shows the comprehensive literature review of various optimization techniques applied for EDM optimization.

1.4   Abrasive Water Jet Machining (AWJM)

In a generalized Abrasive Jet Machining process very high velocity fine abrasive particles impinge on the work material resulting in cutting of the workpiece. This is very effective method for hard and brittle materials [19]. Abrasive Water Jet Machining AWJM is an advanced version of AJM which employs water as the carrier medium for abrasive particles. The AWJM process can machine complex shapes and importantly, doesn't generate heat concentrated zones [24]. It can machine a wide variety of materials including advanced layered composites. Further development of abrasives leads to AWJM capability improvement [1]. Figure 1.3 illustrates the AWJM process.

Table 1.2 below shows the comprehensive literature review of various optimization techniques applied for AWJM optimization.

Table 1.1 Comprehensive review of the algorithms/strategies used for EDM process

Reference	Problem addressed	Algorithm	Improvement	Weakness/limitations	Tools/datasets
Das et al. [7]	MRR and surface roughness R_a	RSM, ABC	Simulation results were in good agreement with experimental values	ABC algorithm solutions were compared with experimental values only	machining parameters: pulse on-off time, discharge current and voltage
Bhattacharya et al. [4]	Minimize surface roughness R_a, white layer thickness and surface crack density	RSM	Mathematical models proved to be accurate and interactive.	Only two process parameters: peak current and pulse on duration have been considered	M2 Die Steel and SEM images, detailed spectrograph
Yang et al. [36]	Maximize MRR and minimizing surface roughness R_a	CPNN, SA	Constrained multi-objective optimization methodology validated	The effect of Pulse duration and peak current were not evaluated	ANN model developed based on experimental data
Gopalakannan and Senthilvelan [14]	Maximize MRR and minimize surface roughness R_a and electrode wear rate	RSM, ANOVA	Aluminium metal matrix composite prepared by liquid metallurgy process	Solutions were compared with experimental values only	30 experiments with process parameters: pulse current, gap voltage, pulse on-off time
Teimouri and Baseri [35]	Maximize MRR and minimizing surface roughness R_a	ANFIS, CACO	Hybrid ANFIS and CACO yielded improved performance	Interactive effect of Pulse current, gap voltage, pulse on-off time were not evaluated	Process parameters: magnetic field intensity, rotational speed of electrode, discharge current and pulse on-time
Dewangan et al. [8]	Improve surface integrity aspects	RSM, GRA, Fuzzy Logic	Hybrid GRA and fuzzy-logic yielded improved performance	Limited number of experiments and comparison	Process parameters: Discharge current, pulse-on time, tool-work time and tool-lift time

(continued)

Table 1.1 (continued)

Reference	Problem addressed	Algorithm	Improvement	Weakness/limitations	Tools/datasets
Kolli and Kumar [21]	Optimize surfactant and graphite powder concentration in dielectric fluid maximize MRR and minimizing surface roughness R_a, tool wear rate, recast layer thickness	Taguchi method, ANOVA	Significant improvement in MRR, and surface roughness R_a, tool wear rate, and recast layer thickness	The process parameters such as pulse on time, gap voltage, pulse off time were not evaluated	Process parameters: discharge current, surfactant concentration and powder concentration
Camposeco-Negrete [5]	Minimize machining time and surface roughness R_a	Taguchi method, Experimentation, Desirability approach	Reduction in machining time and surface roughness R_a	The approach was limited to experimental approach.	Process parameters: pulse-on-off time, servo voltage, and wire speed
Ganapathy et al. [13]	Maximize MRR and tool wear rate	RSM, ANOVA	Influence of tool size and peak current on MRR and tool wear rate was established	The approach was limited to experimental approach	Process parameters: peak current, pulse on time, dielectric pressure and electrode size
Dang [6]	Maximize MRR, electrode wear rate	Taguchi method, PSO, Kriging modelling	Influence of discharge current on MRR was established	Solutions of PSO have not were compared against experimental solutions only	Process parameters: spark current rate, voltage of spark gap, pulse on-off time

Fig. 1.3 Illustration of AWJM Process [17]

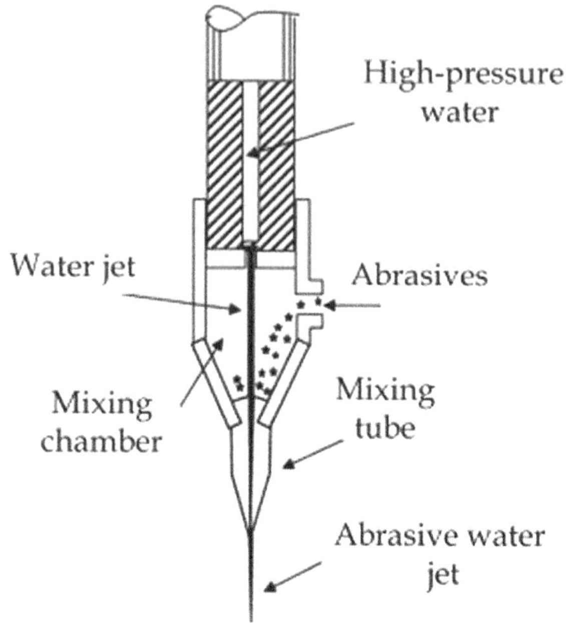

1.5 Tool Based Micro-machining Processes

In recent years, the concept of miniaturization for industrial products is on growing due to developments in the field of aerospace, medical implants, semiconductor industries, automobile field, etc. Micromachining is the key technology as it plays an important role in today's manufacturing field in terms of increased number of functions with reduced dimensions [2]. It is the most basic machining technology for the production of such miniaturized parts and components. Micro-machining based on lithography has several disadvantages as compared to the tool-based micro-machining processes such as micro-turning, micro-milling and micro-drilling, micro-grinding, etc. Tool based micromachining processes have many advantages in productivity, efficiency, flexibility and cost effectiveness. Figure 1.4 describes the classification of micro fabrication processes.

Mechanical micro-machining is a scaled-down version of conventional turning, drilling and milling processes. This process can also be categorized as micro-milling, micro-turning, or micro-drilling depending on the same basic machining principle applied to categorize conventional macro-scale machining processes. Mechanical micro-machining is capable of fabricating complex miniature parts as small as tens of microns. The machining performance is greatly influenced by several parameters such as chip thickness, tool edge radius and micro-structure of workpiece material that affect cutting forces, vibrations, process stability and finally surface finish [10]. It is necessary to understand the process physics for improvement of the accuracy, productivity and wide applicability of this technique to majority of the materials [18].

Table 1.2 Comprehensive review of the algorithms/strategies used for AWJM process

Reference	Problem addressed	Algorithm/strategy	Improvement	Weakness/limitations	Tools/datasets
Armagan and Arici [1]	Minimize surface roughness R_a & $kerf$ angle	Taguchi Method, ANOVA	Standoff distance was found to be the most effective parameter	The approach was limited to experiments	Abrasive mass flow rate, and water pressure
Shanmugam et al. [31]	Minimize $kerf$ angle	Taguchi Method	Improved water pressure, traverse speed and standoff distance	The approach was limited to experiments	Abrasive mass flow rate, and water pressure
Shukla and Singh [33]	Minimizing $kerf$ angle	FA, PSO, BH, ABC, SA, GA and BBO	BBO yielded better solution quality	Parameter setting of the FA needed significant number of preliminary trials.	Mass flow rate
Schwartzentruber et al. [30]	Maximize the abrasive energies	GA	Optimized nozzle designs with faster MRR	The effects of substrate properties and impact angle have not been evaluated	Mass flow rate
Jain et al. [20]	Optimization of USM, AJM, WJM and AWJM processes	GA	Performance of USM, AJM, WJM and AWJM processes improved	GA is compared with experimental values only	Water jet pressure at nozzle exit, diameter of abrasive-water jet nozzle, mass flow rate of water and abrasives mass flow rate

(continued)

Table 1.2 (continued)

Reference	Problem addressed	Algorithm/strategy	Improvement	Weakness/limitations	Tools/datasets
Gostimirovic et al. [16]	Optimization of trajectory curvature	Genetic Programming (GP)	GP yielded effective and robust results	GP is compared with experimental values only	Water pressure, cutting speed and abrasive mass flow rate
Zain et al. [38]	Minimize surface roughness R_a	Regression, SA, GA, integrated SA–GA	SA-GA performed exceedingly better as compared to other techniques	Integrated GA-SA results have been compared with experimental values only	Experimental data

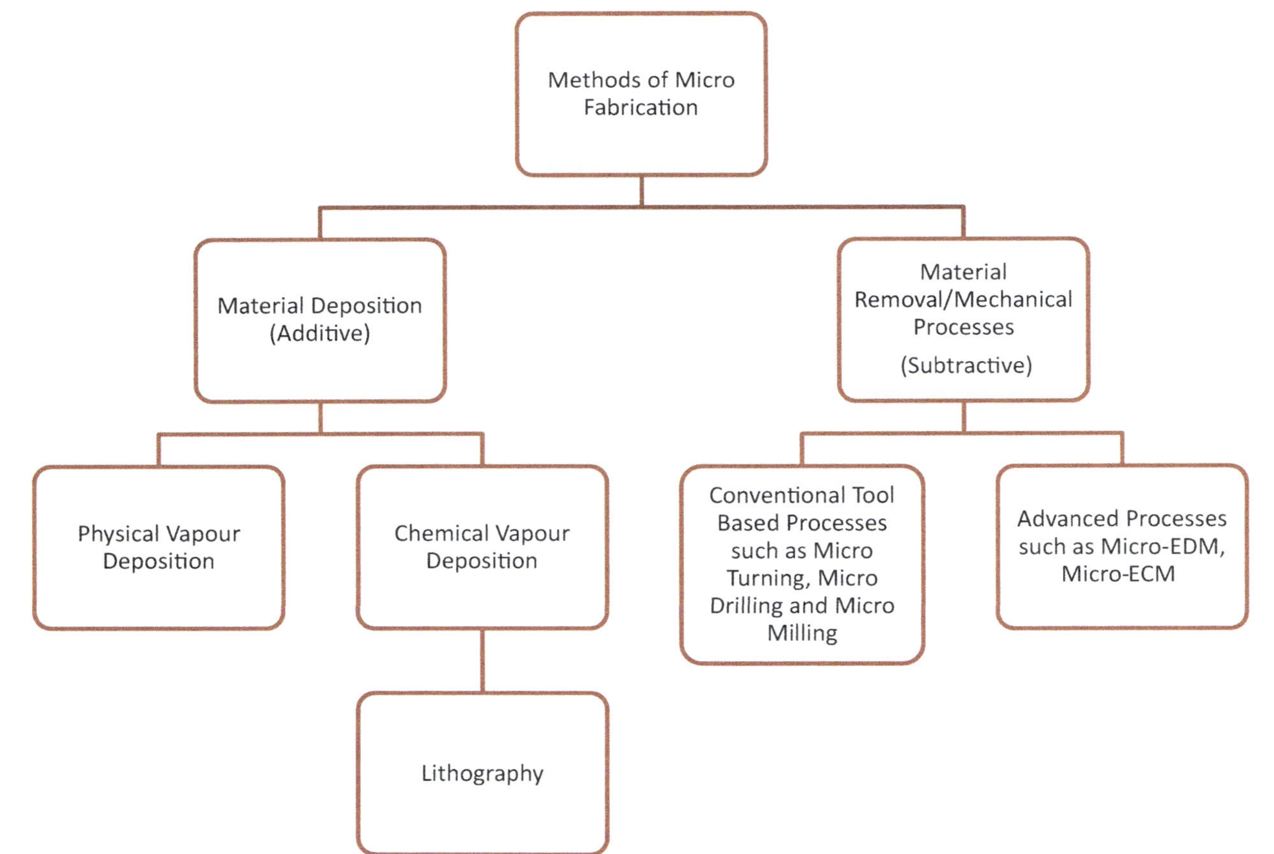

Fig. 1.4 Classification of micro machining processes

The working phenomenon for micro turning is analogous to conventional machining operation. The material is removed from the workpiece by means of micro tools, however, it has certain characteristics due to its significant size reduction. In micromachining, the selection of material for a specific application is a challenging task. Since the size is very small, material property plays a major role in functionality of process and final product. In micro turning operation, the accuracy of the finished micro parts is greatly influenced by machining forces generated; hence, the control of the forces is imperative. Mechanical micro drilling is a cost efficient method used for the generation of micro holes in miniature parts. Unlike additive micromachining processes, the electrical property does not influence the micro drilling operation. The micro drilling of brass is shown in Fig. 1.5. Generation of micro holes in printed circuit boards (PCBs) is the best example of the process. Burr formation, chip morphology, effect of Lubrication, cutting tool material and characteristics, wear aspects of tool, process parameter optimization are the focused areas for investigation of micro drilling processes.

Micro milling is a process of end milling with the milling cutter diameters ranging from 0.1 to 1 mm. It is similar to the conventional milling operation, in which the spindle rotates the end mill cutter to remove the material from the workpiece. It is used to machine complex 3D micro parts and micro features in micro and miniature components [12]. The typical applications of micro milling includes micro moulds, micro dies, micro channels, micro gears, micro fluidic devices, micro propellers, and micro heat exchangers [23]. End mill cutters, ball nose tools, and engraving tools are commonly used micro tools for micro milling operations. These tools are made up of tungsten carbide and hence possess relatively high hardness and strength at elevated temperatures. Due to this, difficulties associated with other micro manufacturing

Fig. 1.5 Micro drilling of brass [29]

Fig. 1.6   Micro milling of PMMA with 0.7 mm micro cutter [22]

processes such as low material removal rate, burning effect, and long throughput time have been overcome by micro milling process [23]. Micro milling experimental setup using 0.7 mm carbide tool for poly methyl methacrylic (PMMA) material is shown in Fig. 1.6. The same problem has been considered in the book for the optimization of micro milling process (Table 1.3).

1.6   Summary

In this chapter, a detailed discussion regarding the concepts and optimization of Advanced Manufacturing Processes (AMPs) such as EDM, AWJM, and Micro Machining has been presented. At the beginning, the concept and classification of AMPs has been discussed. Further, the need for the mathematical modelling and optimization has been illustrated. AMPs such as AWJM, EDM, and Micro Machining processes viz. Micro Turning, Micro Drilling, and Micro Milling have been discussed thoroughly with the schematics of the processes. Moreover, a detailed comprehensive literature review for algorithms/strategies used for each of the process has also been presented.

Table 1.3 Comprehensive review of the algorithms/strategies used for micromachining process

Reference	Problem Addressed	Algorithm	Strategy/Improvement	Weakness/Limitations	Tools/Datasets
Durairaj and Gowri [11]	Minimize surface roughness R_a and tool wear	Regression, GA	Conflicting objectives: tool wear and surface roughness R_a simultaneously optimized	Solutions of GA compared with experimental results only	Experimentation parameter values: cutting speed, feed rate, depth of cut
Sofuoğlu [34]	Optimize Surface roughness R_a prediction	ANN, CART, SVM, Regression	Improved computational cost	Results obtained for proposed mathematical model only.	Experimental data: conventional, hot ultrasonic assisted turning operations
Palani et al. [26]	Optimize surface roughness R_a, tool wear ratio and MRR	ANFIS, Machine Vision System	Predicted values were in agreement with experimental values	ANFIS solutions compared with experimental results only	Experimental: ANFIS training, test data sets, three level full factorial design
Rehman et al. [27]	Failure mechanism of tool and machining parameters affecting micro end milling	Micro milling experimentation	Tool failure reasons investigated	High cutting speed was not considered	Cutting speed: 8000–29,000 rpm, 1 mm diameter milling cutter
Gopalasamy et al. [15]	Experimental investigation of Micro milling parameters	Taguchi method, ANOVA, Regression	Most influential parameters investigated	Limited data used for better designing of the experiments	Total 18 experiments performed for Taguchi and regression model
Zain et al. [37]	Minimize surface roughness R_a	GA, Regression	Improved GA performance was validated	GA compared with experimental results only	Total 24 experiments designed for Taguchi method

(continued)

Table 1.3 (continued)

Reference	Problem Addressed	Algorithm	Strategy/Improvement	Weakness/Limitations	Tools/Datasets
Kumar et al. [22]	Minimize surface roughness R_a, machining time	GA, RSM, Regression	The optimum process parameters improved machining time	Limited cutting speed: 1500–2500 rpm	Experiments with milling cutters of diameters 0.7 and 1 mm
Bao et al. [3]	Minimize surface roughness R_a, machining time	GONNS, GA	Best possible compromise between roughness and machining time was attained	Very high computational cost	Experimental values: surface roughness R_a, machining time
Rahman et al. [27]	Optimize MRR, surface roughness R_a, dimensional accuracy and burr thickness	Micro drilling experiments	Influence of tool diameter on the value of surface roughness was validated	Very limited experiments were performed	Micro drilling experiments with 0.5–1.0 mm drill, Spindle speed and feed at three levels

References

1. Armağan M, Arici AA (2017) Cutting performance of glass-vinyl ester composite by abrasivewater jet. Mater Manuf Processes 32(15):1715–1722
2. Asad ABMA, Masaki T, Rahman M, Lim HS, Wong YS (2007) Tool-based micro-machining. J Mater Process Technol 192:204–211
3. Bao W, Chen P, Tansel I, Reen NS, Yang S, Rincon D (2003) Selection of optimal cutting conditions by using the genetically optimized neural network system (GONNS). In: Kaynak O, Alpaydin E, Oja E, Xu L (eds) Artificial neural networks and neural information processing—ICANN/ICONIP 2003. ICANN 2003, ICONIP 2003. Lecture Notes in Computer Science, vol 2714. Springer, Berlin, Heidelberg
4. Bhattacharyya B, Gangopadhyay S, Sarkar BR (2007) Modelling and analysis of EDMed job surface integrity. J Mater Process Technol 189(1–3):169–177
5. Camposeco-Negrete C (2019) Prediction and optimization of machining time and surface roughness of AISI O1 tool steel in wire-cut EDM using robust design and desirability approach. Int J Adv Manuf Technol 1–12
6. Dang XP (2018) Constrained multi-objective optimization of EDM process parameters using kriging model and particle swarm algorithm. Mater Manuf Processes 33(4):397–404
7. Das MK, Kumar K, Barman TK, Sahoo P (2014) Application of artificial bee colony algorithm for optimization of MRR and surface roughness in EDM of EN31 tool steel. Procedia Mater Sci 6:741–751
8. Dewangan S, Gangopadhyay S, Biswas CK (2015) Multi-response optimization of surface integrity characteristics of EDM process using grey-fuzzy logic-based hybrid approach. Eng Sci Technol Int J 18(3):361–368
9. Di Orio G, Cândido G, Barata J, Scholze S, Kotte O, Stokic D (2013) Self-learning production systems (SLPS)-optimization of manufacturing process parameters for the shoe industry. In: 2013 11th IEEE international conference on industrial informatics (INDIN). IEEE, pp 386–391
10. Dow TA, Scattergood RO (2003) Mesoscale and microscale manufacturing processes: challenges for materials, fabrication and metrology. In: Proceedings of the ASPE winter topical meeting, vol. 28. pp 14–19)
11. Durairaj M, Gowri S (2013) Parametric optimization for improved tool life and surface finish in micro turning using genetic algorithm. Procedia Eng 64:878–887
12. Filiz S, Ozdoganlar OB (2010) A model for bending, torsional, and axial vibrations of microand macro-drills including actual drill geometry-part 1: model development and numerical solution. J Manuf Sci Eng 132:041017–1–8
13. Ganapathy S, Balasubramanian P, Senthilvelan T, Kumar R (2019) Multi-response optimization of machining parameters in EDM using square-shaped nonferrous electrode. In: Adv Manuf Process. Springer, Singapore, pp 287–295
14. Gopalakannan S, Senthilvelan T (2014) Optimization of machining parameters for EDM operations based on central composite design and desirability approach. J Mech Sci Technol 28(3):1045–1053
15. Gopalsamy BM, Mondal B, Ghosh S (2009) Taguchi method and ANOVA: An approach for process parameters optimization of hard machining while machining hardened steel
16. Gostimirovic M, Pucovsky V, Sekulic M, Rodic D, Pejic V (2019) Evolutionary optimization of jet lag in the abrasive water jet machining. Int J Adv Manuf Technol 101(9–12):3131–3141
17. Gupta K, Jain NK, Laubscher R (2017) Chapter 4-advances in gear manufacturing. In: Advanced gear manufacturing and finishing, 67–125
18. Huo D, Cheng K, Wardle F (2010) Design of a five-axis ultra-precision micro-milling machine—ULTRAMIll. Part 1: holistic design approach, design considerations and specifications. Int J Adv Manuf Technol 47(9–12): 867–877
19. Jagadeesh T (2015) Non traditional machining. Mechanical Engineering Department, National Institute of Technology, Calicut
20. Jain VK (2008) Advanced (non-traditional) machining processes. In: Machining. Springer, London, pp 299–327

21. Kolli M, Kumar A (2015) Effect of dielectric fluid with surfactant and graphite powder on electrical discharge machining of titanium alloy using Taguchi method. Eng Sci Technol Int J 18(4):524–535

22. Kumar SL, Jerald J, Kumanan S, Aniket N (2014) Process parameters optimization for micro end-milling operation for CAPP applications. Neural Comput Appl 25(7–8):1941–1950

23. Malekian M, Park SS, Jun MBG (2009) Modeling of dynamic micro-milling cutting forces. Int J Mach Tools Manuf 49:586–598

24. Momber AW, Kovacevic R (2012) Principles of abrasive water jet machining. Springer Science& Business Media

25. Muthuramalingam T, Mohan B (2015) A review on influence of electrical process parameters in EDM process. Arch Civil Mech Eng 15(1):87–94

26. Palani S, Natarajan U, Chellamalai M (2013) On-line prediction of micro-turning multi-response variables by machine vision system using adaptive neuro-fuzzy inference system (ANFIS). Mach Vis Appl 24(1):19–32

27. Rahman M, Kumar AS, Prakash JRS (2001) Micro milling of pure copper. J Mater Process Technol 116(1):39–43

28. Rahman AA, Mamat A, Wagiman A (2009) Effect of machining parameters on hole quality of micro drilling for brass. Modern Appl Sci 3(5):221–230

29. Rao RV (2010) Advanced modelling and optimization of manufacturing processes: international research and development. Springer Science & Business Media

30. Schwartzentruber J, Narayanan C, Papini M, Liu HT (2016) Optimized abrasive waterjet nozzle design using genetic algorithms. In: The 23rd international conference on water jetting, At Seattle, USA

31. Shanmugam DK, Nguyen T, Wang J (2008) A study of delamination on graphite/epoxy composites in abrasive waterjet machining. Compos A Appl Sci Manuf 39(6):923–929

32. Shastri AS, Nargundkar A, Kulkarni AJ (2020) Multi-cohort intelligence algorithm for solving advanced manufacturing process problems. Neural Comput Appl. https://doi.org/10.1007/s00521-020-04858-y

33. Shukla R, Singh D (2017) Experimentation investigation of abrasive water jet machining parameters using Taguchi and Evolutionary optimization techniques. SwarmEvolut Comput 32:167–183

34. Sofuoğlu MA, Çakır FH, Kuşhan MC, Orak S (2019) Optimization of different non-traditional turning processes using soft computing methods. Soft Comput 23(13):5213–5231

35. Teimouri R, Baseri H (2014) Optimization of magnetic field assisted EDM using the continuous ACO algorithm. Appl Soft Comput 14:381–389

36. Yang XS, Deb S (2009) Cuckoo search via Lévy flights. In: 2009 world congress on nature & biologically inspired computing (NaBIC). IEEE, pp 210–214

37. Zain AM, Haron H, Sharif S (2010) Application of GA to optimize cutting conditions for minimizing surface roughness in end milling machining process. Expert Syst Appl 37(6):4650–4659

38. Zain AM, Haron H, Sharif S (2011) Optimization of process parameters in the abrasive waterjet machining using integrated SA–GA. Appl Soft Comput 11(8):5350–5359

Chapter 2
A Brief Review of Socio-inspired Metaheuristics

There are several deterministic and approximation algorithms proposed so far. As the problem complexity grows the approximation algorithms have proven to be computationally cheaper as compared to the earlier ones. The approximation algorithms could be classified as bio-inspired algorithms, swarm-based algorithms and physical & chemical based algorithms. The notable bio-inspired algorithms are Genetic Algorithms, Differential Evolution, Artificial Immune System, etc. The algorithms such as Ant Colony Optimization, Particle Swarm Optimization, Firefly Algorithm, etc. are swarm-based algorithms. According to Kumar et al. [19], the socio-cultural inspired optimization methods are based on the intelligence exhibited by the group of individuals in the framework of socio-political ideologies, competitive behavior in sports, cultural interactions in society, colonization, etc. The individuals in such societal framework compete and interact with one another to become the best within. This further helps in the evolution of entire society. This is the basis of the socio-cultural optimization algorithms as well. So far, the algorithms like Ideology Algorithm (IA) [8], Election Algorithm (ELA) [4], and Election Campaign Algorithm (ECO) [23] have been proposed from socio-political ideologies framework. The League Championship Algorithm (LCA) Kashan [5–10], Soccer League Competition Algorithm (SLC) [24, 25] have been proposed from the framework of competitive behavior in sports. The algorithms such as Teaching Learning Based Optimization (TLBO) [28], Cultural Evolution Algorithm (CEA) [20], Social Learning Optimization (SLO) [21] and Cohort Intelligence (CI) [16] have been proposed from the social and cultural Interaction. The Society and Civilization Algorithm (SCO) [29], Imperialist Competitive Algorithm (ICA) [2, 7], Anarchic Society Optimization (ASO) [1] are from within the societal competition in the colonization. These methods are becoming popular as they are based on simple rules and can handle a wide variety of problems from continuous, discrete and combinatorial domain. A detailed classification of

A. Shastri et al., *Socio-Inspired Optimization Methods for Advanced Manufacturing Processes*, Springer Series in Advanced Manufacturing, https://doi.org/10.1007/978-981-15-7797-0_2

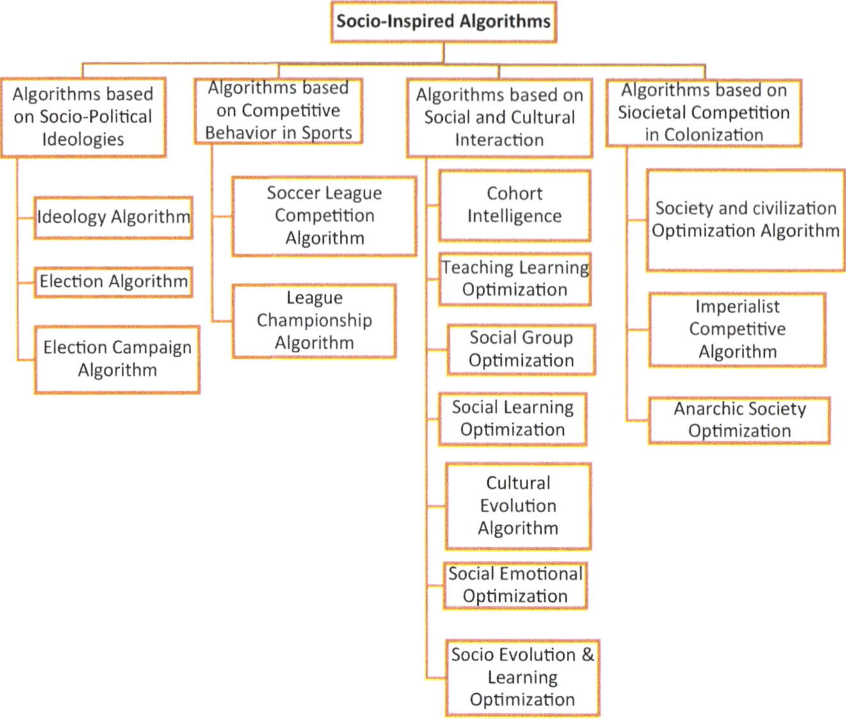

Fig. 2.1 The socio-inspired algorithms [19]

socio inspired algorithms is given in Fig. 2.1. These algorithms have been reviewed in brief in the next section.

2.1 Algorithms Based on Socio-Political Ideologies

The IA algorithm is inspired from the prevalent ideologies in the political domain which divides the society. The individuals inspired from certain ideology supports and contributes to the party to win as well as exhibit self-interested behavior to raise in the ladder of the progress. These individuals also motivate from other parties to follow them. Every party individual observes the local part leader and the other party leaders and the comparison further helps it to decide its own path. Furthermore, every local party leader desire to be a global leader and competes with the peers in the own party as well as the leader from the other parties. In addition, the lowest rank individual following the party ideology desires to climb up in the party, however, for prolonged time if it understands that following its current party ideology is not improving its rank. Such deserted individual may change the ideology and resort to another party ideology. This competition amongst individuals within the political parties as well as within the individuals from within different competing parties is modeled in IA. The algorithm is validated Wilcoxon signed-rank test by comparing against contemporary metaheuristics.

ELA is inspired from the elections between competing political parties. An election is a mechanism of democratically selecting a public representative. The algorithm models the entire election process from campaigning, positive and negative advertisement, coalition and winning of the leader along with the population consisting of voters and candidates. The algorithm forms the political parties by grouping the individuals based on the closeness of Euclidean distance. These individuals then positively advertise for its own party and negatively advertise for the other parties. In this iterative process, the disliked candidates loose out and stuck in the local minima; however, the liked candidates search the solution further to the convergence. As the multiple political parties with similar beliefs form a coalition party, the algorithm also collates two or more solutions when in close neighborhood of one another. A candidate is then chosen amongst the coalition party individuals as the leader and rest of the candidates and voters become the followers. The algorithm is assumed to have converged when a candidate secures highest number of votes and represents final or optimal solution. The performance of the algorithm was validated by comparing against several existing algorithms.

The ECO algorithm models the mannerisms of political candidates during an election campaign. The solution space is divided amongst the voters and the electoral candidates. The feasible solutions are referred to as the electoral candidates. The candidates influence the voters based on their prestige (solution value). The votes are based on the prestige and finally a stronger election candidate wins the highest votes. The prestige of the candidates is updated iteratively till the convergence and winning of a single candidate. The algorithm so far has been applied for applied test functions and dynamic control problems.

2.2 Algorithms Based on Competitive Behavior in Sports

LCA models the competition amongst teams in the league matches. The modeled league teams (representing solutions) compete over a period of few weeks and the strong teams emerge out of them. At the end the strongest team is considered as winner. The weeks represent iterations, teams represent solutions and the end of the playing season is considered as stopping criterion, the strength of the team is considered as the fitness of the solution. In this championship algorithm the teams, strength, schedule of the season is initialized randomly based on the pre-decided league size. The schedule is decided in a round-robin manner in which every team contests every other team once in the season. Team members or individuals from sport teams then compete with each other across weekly matches based on the match schedule. The matches are played with a pair of teams based on a defined weekly league schedule. As every team represents a solution, the strength of the solution is referred to a fitness of the solution. The strategies are revised based on the current solutions of all the teams and the session is iteratively repeated. This process finally evolves into a best fit team, i.e. a best solution. So far, the LCA has been applied for solving mechanical design problems as well as several constrained problems.

The competitive behavior amongst the teams and players in soccer league matches is modeled in SLC algorithm. Various soccer sporting clubs compete for the league series where every team competes twice with every team. The teams are ranked on the basis of their weekly wins and the sporting club with the maximum score at the end of the season is declared as the season champion. The SLC models every individual team member as a solution vector. The associated objective function solution is referred to as power of the team member. So, the fixed and substitute members of every team constitute the population of solutions. The total power of the team is computed as average power value of all of its players. The team with the higher power carries higher probability of winning the match. Along with this competition amongst team members, every player competes with every other member to become star player (local solution) or super star player (global solution). Similar to the mutation in GA, every solution is changed for possible local improvement. The associated important parameters are imitation operator, provocation operator, mutation operator, and substitution operator. The imitation operator helps the players to follow behavior of a star player from their own team and a super star player. The provocation operators such as forward replacement, backward replacement or a random replacement helps the substitute players to improve their performance and find a permanent position. The solution vectors from the losing teams update their playing behavior and playing positions through the mutation operator which helps in imparting diversification of the solutions. The algorithm continues its iteration till convergence conditions are satisfied. So far, the SLC algorithm has been applied for solving optimal design of water distribution networks, knapsack problem, etc.

The individuals in a society interact with one another to improve their overall behavior and a cooperative interaction among such societies represents a civilization. The SCO is inspired from interactive and competitive human social behavior amongst the civilized society individuals. The algorithm models a group of such individuals as a society and a collection of such societies as a civilization. The fittest individual or solution is represented as a leader which facilitate other individuals to improve. Similar to IA, the individual solutions in the SCO algorithm try to move towards their leaders or fitter individuals in the society and the leaders in the society try to move towards better leaders across the other societies or better promising search spaces. Thus, every individual directly and indirectly improves itself through intra- and inter-societal interactions. This will eventually lead to the improvement in the solution of the individuals as well as society as a whole. As the algorithm progresses there will fewer individuals in the lesser promising areas of the problem search space. It is important to mention that the intra-society learning imparts exploitation capability and migration of leaders towards the better society leaders imparts exploration capability to the algorithm.

2.3 Algorithms Based on Societal Competition in the Colonization

The ICA models socio-political and behavior of imperialist nations which competes to extend their power by gradually acquiring the weaker nations/empires. This competition finally results in strengthening the power of stronger and successful imperialist empire(s) and the weaker empires collapse gradually. Similar to ELA, the initial population in ICA randomly generates initial countries. These countries are considered as candidate solutions or the objective function values. These values are considered as strength of the country. The best countries are chosen iteratively and are considered as the emerging imperialist nations. The imperialist empires are allotted these colonies based on the power of the colonies. The allotment here is referred to as colonies moving towards the corresponding imperialist nation. The revolution operator introduces random changes in the behavior of the countries. This enables the weak countries acquire better position in the local neighborhood and may further help them jump out of local minima. This imperialist competition results in gradually increasing the power of stronger and successful imperialist empires; whilst the weaker empires collapse gradually leading to a state of convergence.

The members of anarchic society randomly and adventurously try to search for the better solution. This is a commonly observed human behavior of being greedy and disorderly to improve or achieve their goals to the best. The ASO models the members of an anarchic society which are fickle minded, irrational and frequently display unruly behavior. The members attempt to find the best global best solution in the unknown search space. As the algorithm progresses each member computes a new position to move based on his own individual understanding. Each member determines its behavior or next position in the search space based on its own and others previous experiences. The first movement strategy is based on its current position, where the anarchic member computes a fickleness index to measure the satisfaction of his current position in comparison others. A small fickleness index indicates that the current member is placed best in comparison to other members; otherwise the current member chooses to make a fickle and an unpredictable movement. In the second movement policy, the member computes an external irregularity index based on the global best position of other members during the earlier iterations. The member behaves anarchically if its behavior (fitness) is located away from the global best; else it exhibits a logical behavior. The third movement policy is based on the individual's past positions (his previous best). For this, the member computes an internal irregularity index and then displays anarchic behavior (unpredictable random movement) or behaves logically, i.e. moves to a position closer to its previous best). The resulting and the final movement is done either as elitist, sequential or through combinational rules. Under the elitist method, out of the three movement policies, the one with the best answer is chosen as the resulting movement or the movement policies may be combined sequentially, or the moves may be accepted as a combination rule. Thus, based on the chosen movement policy, the current member updates its current situation. Similarly, the entire member updates their positions (fitness) and the algorithm continues till convergence.

2.4 Algorithms Based on Social and Cultural Interaction

The TLBO algorithm is inspired from the learning process involving teacher and students. In the general education environment, the two essential phases are, teaching and learning. The learning is done through the teaching by the teacher and peers. If the teacher is better along with the competitive environment, then the students can become better. These phases and the associated improvement in the competitive environment have been modeled in TLBO in which a group of learners represent the initial population. Every modeled student has several subjects to be learnt. These subjects are considered as variables to be optimized. The algorithm begins with the initialization of a population of students. Every student computes the mean of all the subjects, which represents fitness of the computed solution. The best solution amongst the population is considered as the teacher. The students then improve their variable values by following the teacher as well as by following the peers. This helps in exploration as well as exploitation. The process continues till convergence.

CEA is inspired from the continual changes in culture evolving the society towards survival. The collaborative efforts amongst the societal individuals are at the heart of the cultural evolution. These efforts are through the means of group consensus, individual learning, innovative learning and self-improvement. The CEA models these means of cultural evolution. In the group consensus model, the characteristics of the whole cultural population are indexed and based on it weaker species can learn and improve their solutions. In the innovative learning model, the learning of certain characteristics from existing species helps in achieving diversified solutions. The self-improvement is basically searching for better solution in the close neighborhood of the existing solution. In the CEA environment the initial cultural population is generated randomly and ranked based on the fitness value. Similar to GA, the elitist species are reserved while the rest go through any of the collaborative efforts discussed above. The modified or culturally evolved species are then compared with one another in terms of their fitness and forces the weaker to die out. This process continues till the pre-decided innovations are reached. A new generation of cultural population is then formed by merging the elitist cultural species. This process continues till convergence and the best cultural species is chosen as the optimal solution.

Genetic and belief based cultural evolution are at the heart of social learning. The cultural evolution influences the genetic evolution and affects the evolution of human intelligence. The SLO models the cultural or societal evolution of the society. The algorithm begins with randomly generated population and then similar to GA the population is evolved using mutation, crossover and the selection operations. This genetic evolution phase continues till the termination conditions are satisfied.

Using the observation learning and imitation operators the individuals exhibit societal evolution. So far, the algorithm has been applied for solving NP-hard discrete optimization problems [21, 22].

CI [16, 13] a successful Socio-inspired metaheuristic mimics the self-learning behavior exhibited by candidates in a group, where the candidates cooperate and compete with one another to achieve some individual goal. A cohort refers to a set of homogeneous agents who coexist as a group to try learning and improving their behavior in the process. The candidates in the cohort interact and every candidate may follow the behavior of another candidate from the cohort which could result in the improvement of its own behavior. The CI algorithm starts with the initialization of the parameters for the algorithm and like other population-based algorithms; the initial population is generated randomly. Every candidate in the population is represented using his qualities (the problem variables) and the associated behavior of that candidate (solution vector or the objective function). The iterative process begins with every candidate choosing to follow a better behaving candidate. This probabilistic choice of which candidate behavior is to be followed or imitated is simulated using the roulette wheel selection approach. It gives fair opportunity to every behavior in the cohort to get selected. Once a candidate chooses to follow the behavior of a particular candidate from the cohort, it updates its qualities in the close neighborhood of the candidate. This represents imitating an individual's qualities and its respective behavior in real life; which may result in an improved and updated behavior of the candidate. This process continues till behavior of the cohort saturates i.e. no significant improvement is observed in the behaviors of the candidates across the cohort. The algorithm then terminates and the best behavior (problem solution) is reported as the behavior of the cohort. The strength of the algorithm lies in its ability to jump out of the local minima. CI achieves this through the follow mechanism where, when a candidate chooses to follow another candidate, it imitates its qualities (the variables). This imitation means that the qualities are not replicated directly; but the candidate generates a set of qualities in the close neighborhood of the candidate's qualities it follows. This introduces possibility of more variations in the solutions generated across the iterations; thus, avoiding getting trapped in the local minima. A generalized CI algorithm flowchart is presented in Fig. 2.2.

In the context of following one another, in addition to roulette wheel selection approach, six independent variations have been proposed which represent different variations of the CI method. They are as follows [27]:

1. Follow Best Rule: Every candidate follows the best candidate in the cohort
2. Follow Better Rule: Every candidate follows next better candidate, and the best candidate follows itself
3. Follow Worst Rule: Every candidate follows the worst candidate in the cohort
4. Follow Itself Rule: Every candidate follows itself
5. Follow Median Rule: Every candidate follows a candidate which has median probability
6. Alienation-and-Random Selection Rule: Every candidate alienates a randomly chosen candidate in the cohort, i.e., it does not follow a particular candidate at

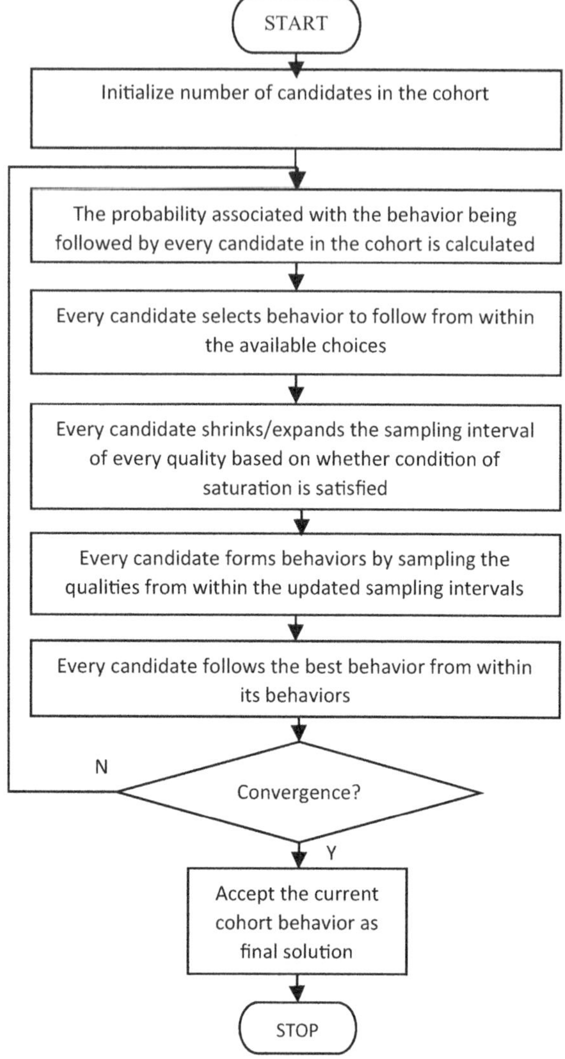

Fig. 2.2 Generalized Cohort intelligence (CI) algorithm flowchart

all. It randomly chooses the candidate to follow from the rest of the candidates including itself.

Sample Code of Roulette Wheel variation is given in the appendix. All other executable codes for all the chapters are given on the following website: https://sites. google.com/site/oatresearch. These variations were tested on certain multimodal and uni-modal problems to establish the suitability of a variation to a specific problem type and a multitude of real-world problems. Further these variations have been successfully applied for optimization of Abrasive Water Jet Machining (AWJM) process [6]. It has been successfully applied to the domain of combinatorial problems like optimizing in image steganography [31, 32], hybridized K-means based data clustering [12], the 0–1 knapsack problem [15], a cyclic bottleneck assignment problem modeled to the problems in health care systems and inventory management [17]. Similarly, it is also applied to solve a multi-dimension multiple-knapsack problem (as a sea cargo mix problem) and to optimizing in the cross-border shipper and cargo assignments problems subject to various constraints [13]. Further, CI has been combined with image processing technique and has been applied to solve the problem of micro drilling machining process optimization [26].

In the domain of discrete problems, CI is applied to the problems of mesh smoothing of complex geometry [30] and to the optimization of shell-and-tube heat exchanger [3]. A version of CI integrated with probability-based constrained handling approach is proposed in [33]. The article evaluated this version of CI by applying it to solve some inequality based constrained problems and analyzed on the robustness and the rate of convergence of this version. Furthermore, CI with static penalty function approach (SCI) and CI with dynamic penalty function approach (DCI) are discussed in [14]. The efficacy of the two proposed constrained handling methods were then evaluated by solving several constrained test problems. CI along with analytical hierarchy process (AHP) was effectively applied to a practical problem of recommending ice-creams to diabetics [5], the results obtained were further compared to an AHP-genetic algorithm version. The successful application of CI to a wide domain of real-world problems points to the capability of CI in handling complex constraints. For detailed mathematical formulation of the CI algorithm please refer to Kulkarni et al. [18].

References

1. Ahmadi-Javid A, Hooshangi-Tabrizi P (2017) Integrating employee timetabling with scheduling of machines and transporters in a job-shop environment: a mathematical formulation and an anarchic society optimization algorithm. Comput Oper Res 84:73–91
2. Atashpaz-Gargari E, Lucas C (2007) Imperialist competitive algorithm: an algorithm for optimization inspired by imperialistic competition. In: 2007 IEEE congress on evolutionary computation. Singapore, pp 4661–4667
3. Dhavle SV, Kulkarni AJ, Shastri A, Kale IR (2018) Design and economic optimization of shell-and-tube heat exchanger using cohort intelligence algorithm. Neural Comput Appl 30(1):111–125
4. Emami H, Derakhshan F (2015) Election algorithm: a new socio-politically inspired strategy. AI Commun 28(3):591–603
5. Gaikwad SM, Joshi RR, Kulkarni AJ (2015) Cohort intelligence and genetic algorithm along with AHP to recommend an ice cream to a diabetic patient. In: Proceedings of the international conference on swarm, evolutionary and Memetic computing. Springer, Cham, pp 40–49
6. Gulia V, Nargundkar A (2019) Optimization of process parameters of abrasive water jet machining using variations of cohort intelligence (CI). In: Applications of artificial intelligence techniques in engineering. Springer, Singapore, pp. 467–474

7. Hosseini S, Al Khaled A (2014) A survey on the imperialist competitive algorithm metaheuristic: implementation in engineering domain and directions for future research. Appl Soft Comput 24:1078–1094

8. Huan TT, Kulkarni AJ, Kanesan J, Huang CJ, Abraham A (2017) Ideology algorithm: a socio-inspired optimization methodology. Neural Comput Appl 28(1):845–876

9. Kashan AH (2011) An efficient algorithm for constrained global optimization and application to mechanical engineering design: league championship algorithm (LCA). Comput Aided Des 43(12):1769–1792

10. Kashan AH (2014) League championship algorithm (LCA): an algorithm for global optimization inspired by sport championships. Appl Soft Comput 16:171–200

11. Kashan AH (2009): League championship algorithm: a new algorithm for numerical function optimization. In: IEEE international conference of soft computing and pattern recognition. Malacca, Malaysia, pp 43–48

12. Krishnasamy G, Kulkarni AJ, Paramesran R (2014) A hybrid approach for data clustering based on modified Cohort intelligence and K-means. Expert Syst Appl 41(13):6009–6016

13. Kulkarni AJ, Baki MF, Chaouch BA (2016) Application of the Cohort-intelligence optimization method to three selected combinatorial optimization problems. Eur J Oper Res 250(2):427–447

14. Kulkarni O, Kulkarni N, Kulkarni AJ, Kakandikar G (2018) Constrained Cohort intelligence using static and dynamic penalty function approach for mechanical components design. Int J Parallel Emergent Distrib Syst 33(6):570–588

15. Kulkarni AJ, Shabir H (2016) Solving 0–1 Knapsack problem using cohort intelligence algorithm. Int J Mach Learn Cybern 7(3):427–441

16. Kulkarni AJ, Durugkar IP, Kumar M (2013) Cohort intelligence: a self supervised learning behavior. In: IEEE international conference on systems, man, and cybernetics (SMC). Manchester, UK, pp 1396–1400

17. Kulkarni AJ, Krishnasamy G, Abraham A (2017) Cohort intelligence: a socio-inspired optimization method. In: Intelligent Systems Reference Library. Springer International Publishing, Switzerland, 114,1–134

18. Kulkarni AJ, Singh PK, Satapathy SC, Husseinzadeh Kashan A, Tai K (eds) (2019) Socio-cultural inspired metaheuristics. In: Studies in computational intelligence, vol. 828. Springer, Singapore

19. Kumar M, Kulkarni AJ, Satapathy SC (2018) Socio evolution & learning optimization algorithm: a socio-inspired optimization methodology. Fut Gener Comput Syst 81:252–272

20. Kuo HC, Lin CH (2013) Cultural evolution algorithm for global optimizations and its applications. J Appl Res Technol 11(4):510–522

21. Liu ZZ, Chu DH, Song C, Xue X, Lu BY (2016) Social learning optimization (SLO) algorithm paradigm and its application in QoS-aware cloud service composition. Inf Sci 326:315–333

22. Liu ZZ, Qin JX, Song C (2017) (2017) Social learning optimization algorithm for functions optimization. J Chinese Comput Syst 38(5):1063–1069

23. Lv W, He C, Li D, Cheng S, Luo S, Zhang X (2010) Election campaign optimization algorithm. Procedia Comput Sci 1(1):1377–1386

24. Moosavian N (2015) Soccer league competition algorithm for solving Knapsack problems. Swarm Evolut Comput 20:14–22

25. Moosavian N, Roodsari BK (2014) Soccer league competition algorithm: a novel Meta-Heuristic algorithm for optimal design of water distribution networks. Swarm Evolution Comput 17:14–24

26. Pansari S, Mathew A, Nargundkar A (2019. An investigation of Burr formation and cutting parameter optimization in micro-drilling of Brass C-360 using image processing. In: Proceedings of the 2nd international conference on data engineering and communication technology. Springer, Singapore, pp 289–302

27. Patankar NS, Kulkarni AJ (2018) Variations of cohort intelligence. Soft Comput 22(6):1731–1747

28. Rao RV, Savsani VJ, Vakharia DP (2011) Teaching–learning-based optimization: a novel method for constrained mechanical design optimization problems. Comput Aided Des 43(3):303–315

29. Ray T, Liew KM (2003) Society and civilization: an optimization algorithm based on the simulation of social behavior. IEEE Trans Evol Comput 7(4):386–396
30. Sapre MS, Kulkarni AJ, Chettiar L, Deshpande I, Piprikar B (2018) Mesh smoothing of complex geometry using variations of Cohort intelligence algorithm. Evoluti Intell 1–16
31. Sarmah DK, Kulkarni AJ (2017) Image steganography capacity improvement using Cohort intelligence and modified multi-random start local search methods. Arabian J Sci Eng 43(8):3927–3950
32. Sarmah DK, Kulkarni AJ (2018) JPEG based steganography methods using Cohort intelligence with cognitive computing and modified multi random start local search optimization algorithms. Inf Sci 430:378–396
33. Shastri AS, Jadhav PS, Kulkarni AJ, Abraham A (2016) Solution to constrained test problems using Cohort intelligence algorithm. In: Innovations in bio-inspired computing and applications. Springer, Cham, pp 427–435

Chapter 3
Multi Cohort Intelligence Algorithm

Multi-Cohort Intelligence (Multi-CI) algorithm has been proposed by Shastri and Kulkarni in [14]. The algorithm implements intra-group and inter-group learning mechanisms. It focuses on the interaction amongst different cohorts. The performance of the algorithm was validated by solving 75 unconstrained test problems with dimensions up to 30. The solutions were comparing with several recent algorithms such as Particle Swarm Optimization, Covariance Matrix Adaptation Evolution Strategy, Artificial Bee Colony, Self-adaptive differential evolution algorithm, Comprehensive Learning Particle Swarm Optimization, Backtracking Search Optimization Algorithm and Ideology Algorithm. The Wilcoxon signed rank test was carried out for the statistical analysis and verification of the performance. The proposed Multi-CI outperformed these algorithms in terms of the solution quality including objective function value and computational cost, i.e. computational time and functional evaluations. The prominent features of the Multi-CI algorithm along with the limitations are discussed as well. In addition, an illustrative example is also solved and every detail is provided.

3.1 Introduction

In recent times, several optimization algorithms have been developed so far. The algorithms such as Evolutionary Algorithms (EAs), Genetic Algorithms (GAs), Swarm Optimization (SO) techniques, Physics and biologically based Methods, Hyper-heuristic Methods, etc. are bio-/nature-inspired techniques. These methods have proven their superiority in terms of solution quality and computational time over the traditional (exact) methods for solving a wide variety of problem classes. In agreement with the no-free-lunch theorem, certain modifications and supportive techniques are required to be incorporated into these methods when applying for solving a variety of class of problems. This motivated the researchers to resort to development of new optimization methods.

An Artificial Intelligence (AI) based socio-inspired optimization methodology referred to as Cohort Intelligence (CI) was proposed by Kulkarni et al. in [7]. It is inspired from the interactive and competitive social behaviour of individual candidates in a cohort. Every candidate exhibits self-interested behaviour and tries to improve it by learning from the other candidates in the cohort. The learning refers to following/adopting the qualities associated with the behaviour of the other candidates. The candidates iteratively follow one another based on certain probability and the cohort is considered saturated/converged when no further improvement in the behaviour of any of the candidates is possible for considerable number of attempts. It is important to mention here that in the current version of the candidates learn from the other candidates of the same cohort. As the selection is based on roulette wheel approach it is not necessary that the candidate will follow the best candidate in every learning attempt. Even though this helps the candidates jump out of local minima, learning options are limited as only intra-group learning exists. In the society several cohorts exist which interact and compete with one another which could be referred to as inter-group learning. This makes the candidates learn from the candidates within the cohort as well as the candidates from other cohorts. In the proposed Multi-Cohort Intelligence (Multi-CI) approach intra-group learning and inter-group learning mechanisms were implemented. In the intra-group learning mechanism, every candidate based on roulette wheel approach chooses a behaviour from within its own cohort. Then it samples certain behaviours from within the close neighbourhood of the chosen behaviour. In the inter-group learning mechanism, every candidate based on roulette wheel approach chooses behaviour from within a pool of best behaviours associated with every cohort. Then it chooses the best behaviour by sampling certain number of behaviours from within the close neighbourhood of both behaviours chosen using the intra-group learning and inter-group learning mechanisms. Detailed procedure of Multi-CI algorithm along with mathematical formulation is discussed in the following section.

3.2 Procedure of Multi-CI Algorithm

Consider a general unconstrained optimization problem (in minimization sense) as follows:

$$Minimize\ f(X) = f(x_1, \ldots x_i, \ldots x_N)$$
$$Subject\ to\ \psi_i^{lower} \leq x_i \leq \psi_i^{upper},\ i = 1, \ldots, N \tag{3.1}$$

In the context of Multi-CI the objective function $f(\mathbf{X})$ is considered as the behavior of an individual candidate in each cohort with associated set of qualities $\mathbf{X} = (x_1, \ldots x_i, \ldots, x_N)$.

The procedure begins with initialization of learning attempt counter $l = 1$, and K cohorts with number of candidates C_k associated with every cohort $k, (k = 1, \ldots, K)$. Every candidate $c(c = 1, \ldots, C_k)$, $k = 1, \ldots, K$ randomly generates qualities $\mathbf{X}_k^c = \left(x_{1,k}^c, \ldots x_{i,k}^c, \ldots, x_{N,k}^c\right)$ from within its associated sampling interval $[\psi_i^{lower}, \psi_i^{upper}]$, $i = 1, \ldots, N$. The parameters such as convergence parameter ε, sampling interval reduction factor r, behavior variations T and T_Z are chosen. The algorithm steps are discussed below and the Multi-CI algorithm flowchart is presented in Fig. 3.1.

Step 1 (Evaluation of Behaviors): The pool of objective functions/behaviors of every candidate $c(c = 1, \ldots, C_k)$ associated with every cohort $k(k = 1, \ldots, K)$ could be represented as follows:

$$
\mathbf{F} = \begin{bmatrix} f\left(\mathbf{X}_1^1\right) , \ldots, & f\left(\mathbf{X}_k^1\right) , \ldots, & f\left(\mathbf{X}_K^1\right) \\ \vdots & \vdots & \vdots \\ f\left(\mathbf{X}_1^c\right) , \ldots, & f\left(\mathbf{X}_k^c\right) , \ldots, & f\left(\mathbf{X}_K^c\right) \\ \vdots & \vdots & \vdots \\ f\left(\mathbf{X}_1^{C_1}\right) , \ldots, & f\left(\mathbf{X}_1^{C_k}\right) , \ldots, & f\left(\mathbf{X}_K^{C_K}\right) \end{bmatrix} = \left[\, \boldsymbol{f}_1 \, , \ldots, \, \boldsymbol{f}_k \, , \ldots, \, \boldsymbol{f}_K \, \right]
$$

$$(3.2)$$

Step 2 (Pool Z Formation): The best behavior (objective functions with minimum value) candidate $\hat{c}_k$, $k(k = 1, \ldots, K)$ in each cohort are chosen and kept in separated pool $\mathbf{Z}$ and the associated set of behaviors $\boldsymbol{F}^Z$ is represented as follows:

$$
\boldsymbol{F}^Z = \left[\min(\boldsymbol{f}_1) , \ldots, \min(\boldsymbol{f}_k) , \ldots, \min(\boldsymbol{f}_K) \right] = \left[f\left(\mathbf{X}_1^{\hat{c}_1}\right) , \ldots, f\left(\mathbf{X}_k^{\hat{c}_k}\right) , \ldots, f\left(\mathbf{X}_K^{\hat{c}_K}\right) \right]
$$

$$(3.3)$$

Step 3 (Probability Evaluation 1): The probabilities associated with each candidate except pool $\mathbf{Z}$ candidates $c(c = 1, \ldots, C_k - 1)$, in every cohort $k(k = 1, \ldots, K)$ are calculated as follows:

$$
p_k^c = \frac{1/f\left(\mathbf{X}_k^c\right)}{\sum_{c=1}^{C_k-1} 1/f\left(\mathbf{X}_k^c\right)}
$$

$$(3.4)$$

Step 4 (Formation of Tbehaviors): Using roulette wheel approach every candidate selects a behaviour from within its corresponding cohort (except pool $\mathbf{Z}$ behaviors) and forms T new behaviours by sampling in close neighbourhood of the qualities associated with the selected behaviour qualities. The neighbourhood of a quality $x_{i,k}^c$ associated with the sampling interval $[\psi_i^{c,lower}, \psi_i^{c,upper}]$, $i = 1, \ldots, N$ of the follower candidate $c(c = 1, \ldots, C_k - 1)$, $k(k = 1, \ldots, K)$ is as follows:

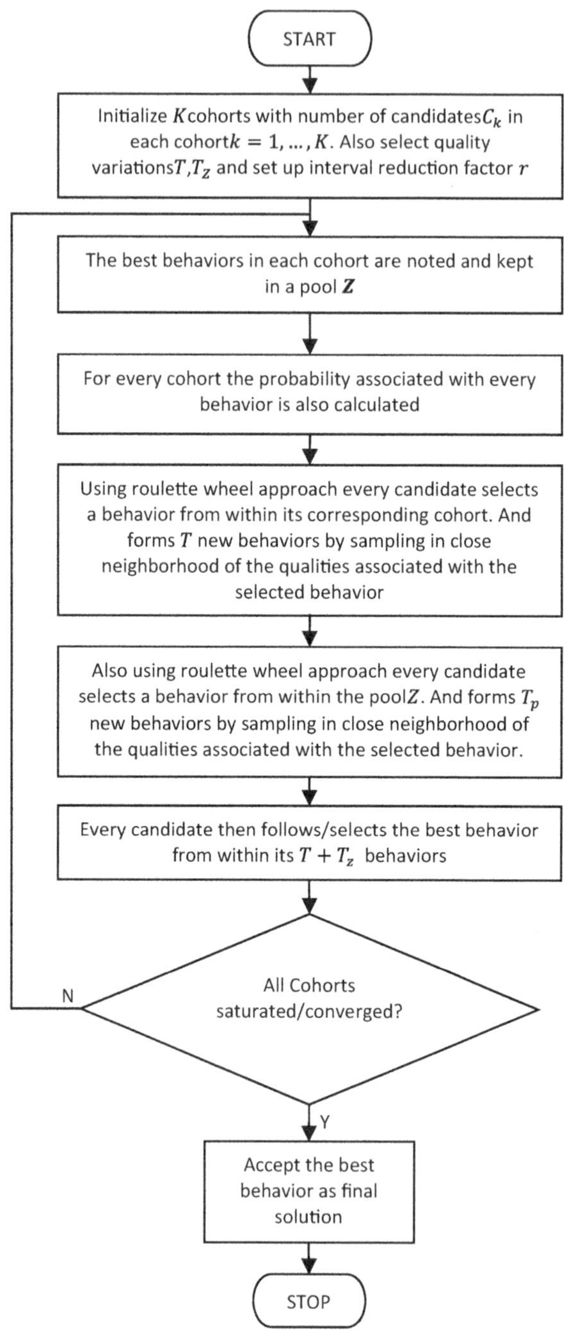

Fig. 3.1 Multi Cohort Intelligence (Multi-CI) flowchart

$$[\psi_i^{c,lower}, \quad \psi_i^{c,upper}] = [x_{i,k}^{\widehat{c}} - \left(\left\|\frac{\psi_i^{\widehat{c},upper} - \psi_i^{\widehat{c},lower}}{2}\right\|\right) \times r, \quad x_{i,k}^{\widehat{c}} + \left(\left\|\frac{\psi_i^{\widehat{c},upper} - \psi_i^{\widehat{c},lower}}{2}\right\|\right) \times r]$$

$$(3.5)$$

where $\widehat{c}$ represents the candidate being followed.

The quality matrix $\mathbf{Z}^T$ associated with every candidate $c(c = 1, \ldots, C_k - 1)$ and corresponding cohort $k(k = 1, \ldots, K)$ is represented as follows:

$$\mathbf{Z}^T = \begin{bmatrix} \mathbf{Z}_1^{1,T} & \cdots & \mathbf{Z}_k^{1,T} & \cdots & \mathbf{Z}_K^{1,T} \\ \vdots & \ddots & \vdots & & \vdots \\ \mathbf{Z}_1^{c,T} & \cdots & \mathbf{Z}_k^{c,T} & \cdots & \mathbf{Z}_K^{c,T} \\ \vdots & & \vdots & \ddots & \vdots \\ \mathbf{Z}_1^{C_1-1,T} & \cdots & \mathbf{Z}_k^{C_k-1,T} & \cdots & \mathbf{Z}_K^{C_K-1,T} \end{bmatrix} \tag{3.6}$$

where $\mathbf{Z}_k^{c,T} = \begin{bmatrix} \mathbf{X}_k^{c,1} \\ \vdots \\ \mathbf{X}_k^{c,t} \\ \vdots \\ \mathbf{X}_k^{c,T} \end{bmatrix} = \begin{bmatrix} x_{1,k}^{c,1} & \cdots & x_{i,k}^{c,1} & \cdots & x_{N,k}^{c,1} \\ \vdots & \ddots & \vdots & & \vdots \\ x_{1,k}^{c,t} & \cdots & x_{i,k}^{c,t} & \cdots & x_{N,k}^{c,t} \\ \vdots & & \vdots & \ddots & \vdots \\ x_{1,k}^{c,T} & \cdots & x_{i,k}^{c,T} & \cdots & x_{N,k}^{c,T} \end{bmatrix}$

The behavior matrix $\mathbf{F}^T$ associated with $\mathbf{Z}^T$ could be represented as follows:

$$\mathbf{F}^T = \begin{bmatrix} f\left(\mathbf{Z}_1^{1,T}\right) & \cdots & f(\mathbf{Z}_k^{1,T}) & \cdots & f(\mathbf{Z}_K^{1,T}) \\ \vdots & \ddots & \vdots & & \vdots \\ f\left(\mathbf{Z}_1^{c,T}\right) & \cdots & f\left(\mathbf{Z}_k^{c,T}\right) & \cdots & f(\mathbf{Z}_K^{c,T}) \\ \vdots & & \vdots & \ddots & \vdots \\ f(\mathbf{Z}_1^{C_1-1,T}) & \cdots & f(\mathbf{Z}_k^{C_k-1,T}) & \cdots & f(\mathbf{Z}_K^{C_K-1,T}) \end{bmatrix} \tag{3.7}$$

where $f\left(\mathbf{Z}_k^{c,T}\right) = \begin{bmatrix} f\left(\mathbf{X}_k^{c,1}\right) \\ \vdots \\ f\left(\mathbf{X}_k^{c,t}\right) \\ \vdots \\ f\left(\mathbf{X}_k^{c,T}\right) \end{bmatrix}$

Step 5 (Probability Evaluation 2): The probabilities associated with each pool $\mathbf{Z}$ candidate $\hat{c}_k$ in every cohort $k(k = 1, \ldots, K)$ are calculated as follows:

$$p^{\hat{c}_k} = \frac{1 \big/ f\left(X_k^{\hat{c}_k}\right)}{\sum_{k=1}^{K} 1 \big/ f\left(X_k^{\hat{c}_k}\right)} \tag{3.8}$$

Step 6 (Formation of T_Z behaviors):Also using roulette wheel approach every candidate selects a behavior from within pool Z and forms T_Z new behaviors by sampling in close neighbourhood of the qualities associated with the selected behaviour. The quality matrix Z^{T_Z} associated with every candidate $\hat{c}_k$, $k(k = 1, \ldots, K)$ is represented as follows:

$$Z^{T_Z} = \left[Z^{\hat{c}_1, T_Z} \ldots Z^{\hat{c}_k, T_Z} \ldots Z^{\hat{c}_K, T_Z} \right] \tag{3.9}$$

$$\text{where } Z^{\hat{c}_k, T_Z} = \begin{bmatrix} X_k^{\hat{c}_k, 1} \\ \vdots \\ X_k^{\hat{c}_k, t_Z} \\ \vdots \\ X_k^{\hat{c}_k, T_Z} \end{bmatrix} = \begin{bmatrix} x_{1,k}^{\hat{c}_k, 1} & \cdots & x_{i,k}^{\hat{c}_k, 1} & \cdots & x_{N,k}^{\hat{c}_K, 1} \\ \vdots & \ddots & \vdots & & \vdots \\ x_{1,k}^{\hat{c}_1, t_Z} & \cdots & x_{i,k}^{\hat{c}_k, t_Z} & \cdots & x_{N,k}^{\hat{c}_K, t_Z} \\ \vdots & & \vdots & \ddots & \vdots \\ x_{1,k}^{\hat{c}_1, T_Z} & \cdots & x_{i,k}^{\hat{c}_k, T_Z} & \cdots & x_{N,k}^{\hat{c}_K, T_Z} \end{bmatrix}$$

The behavior matrix F^{T_Z} associated with Z^{T_Z} could be represented as follows:

$$F^{T_Z} = \left[f\left(Z^{\hat{c}_1, T_Z}\right) \ldots f\left(Z^{\hat{c}_k, T_Z}\right) \ldots f\left(Z^{\hat{c}_K, T_Z}\right) \right] \tag{3.10}$$

$$\text{where } f\left(Z^{\hat{c}_k, T_Z}\right) = \begin{bmatrix} f\left(X_k^{\hat{c}_k, 1}\right) \\ \vdots \\ f\left(X_k^{\hat{c}_k, t_Z}\right) \\ \vdots \\ f\left(X_k^{\hat{c}_k, T_Z}\right) \end{bmatrix}$$

Step 7 (Selection): Every candidate $c(c = 1, \ldots, C_k - 1)$ associated with every cohort $k(k = 1, \ldots, K)$ selects the best behavior, i.e. minimum objective function value from within its behavior choices in F^T and F^{T_Z} as follows:

$$f_{k,min}^{c} = min\left(f\left(Z_k^{c,t}\right), f\left(Z^{\hat{c}_k, t_Z}\right) \right), (c = 1, \ldots, C_k - 1), k(k = 1, \ldots, K) \tag{3.11}$$

Step 8 (Concatenation): The pool Z behaviors F^Z(Eq. (3.3)) are carry forwarded to the subsequent learning attempt. The modified pool of behaviors F is given below.

$$
F = \begin{bmatrix}
f^1_{1,min} , \ldots, f^1_{k,min} , \ldots, f^1_{K,min} \\
\vdots \quad \ddots \quad \vdots \quad \ddots \quad \vdots \\
f^c_{1,min} , \ldots, f^c_{k,min} , \ldots, f^c_{K,min} \\
\vdots \quad \ddots \quad \vdots \quad \ddots \quad \vdots \\
f^{C_1}_{1,min} , \ldots, f^{C_k}_{k,min} , \ldots, f^{C_K}_{K,min}
\end{bmatrix} = \begin{bmatrix} f_1 , \ldots, f_k , \ldots, f_K \end{bmatrix} \quad (3.12)
$$

Step 8(Convergence): The algorithm is assumed to have converged if all of the conditions listed in Eq. (3.13)are satisfied for successive considerable number of learning attempts and accept any of the current behaviors as final solution f^* from within K cohorts.

$$
\left.
\begin{array}{l}
max\left(F^l\right) - max\left(F^{l-1}\right) \leq \varepsilon \\
min\left(F^l\right) - min\left(F^{l-1}\right) \leq \varepsilon \\
max\left(F^l\right) - min\left(F^l\right) \leq \varepsilon
\end{array}
\right\}
\qquad (3.13)
$$

where l is learning attempt counter.

An illustrative example (Sphere function with 2 variables) of the above discussed Multi-CI procedure is given below. It includes every details of first learning attempt followed by evaluation of every step (1–8) is listed in Table 3.1 till convergence.

3.2.1 Illustration of Multi-CI Algorithm

An illustrative example (Sphere function with 2 variables: $inimize \sum_{i=1}^{2} x_i^2$, $Subject to - 5.12 \leq x_i \leq 5.12, i = 1, 2$) of the Multi-CI procedure discussed in Sect. 3.2 is given below and detailed steps(1–8) are explained in Table 3.1 till convergence. The convergence plot is shown in Fig. 3.3. The Multi-CI parameters chosen were as follows: number of cohorts $K = 3$, number of candidates $C^k = 3$, reduction factor value $r = 0.98$, quality variation parameters $T = 2$ and $T_p = 4$, the algorithm stopped when the objective function value is less than 10^{-16}. The 3D plot for Sphere function with 2 variables is shown in Figs. 3.2 and 3.3.

Learning Attempt $l = 1$

$$
X = \begin{bmatrix}
0.4426 & -2.7631 & -4.4698 & 1.2841 & -4.4839 & 4.8435 \\
1.7060 & 4.5039 , & -4.4155 & -0.7989, & -4.0203 & -1.1923 \\
-2.2525 & -1.3291 & -2.4503 & 3.8907 & 0.1308 & -1.8813
\end{bmatrix}
$$

(Step 1, Eq. 3.2): $F = \begin{bmatrix} 7.8304 & 21.6280 & 43.5648 \\ 23.1957 & 20.1344 & 17.5841 \\ 6.8402 & 21.1409 & 3.5564 \end{bmatrix}$

(Step 2, Eq. 3.3): $F^Z = \begin{bmatrix} 6.8402 & 20.1344 & 3.5564 \end{bmatrix}$

Table 3.1 Illustration of Multi-CI algorithm solving Sphere function

Learning attempt (l)	X	F (Step 1, Eq. 3.2)
$l = 2$	$\begin{bmatrix} -0.9715 & -0.1627 & 0.0057 & 0.6932 & -1.6579 & -0.5544 \\ -0.9715 & -0.1627 & 0.0057 & 0.6932 & -2.9836 & -0.8647 \\ -2.2525 & -1.3291 & -4.4155 & -0.7989 & 0.1308 & -1.8813 \end{bmatrix}$	$\begin{bmatrix} 0.9703 & 0.4805 & 3.0562 \\ 0.9703 & 0.4805 & 9.6465 \\ 6.8402 & 20.1344 & 3.5564 \end{bmatrix}$

	F^T (Step 4, Eq. 3.7)	F^{Tz} (Step 6, Eq. 3.10)
	$\begin{bmatrix} 6.6966 \\ 20.3143 \end{bmatrix} \begin{bmatrix} 3.2039 \\ 13.4206 \end{bmatrix} \begin{bmatrix} 17.4818 \\ 20.2246 \end{bmatrix}$ $\begin{bmatrix} 4.6984 \\ 4.1166 \end{bmatrix} \begin{bmatrix} 3.5017 \\ 3.7385 \end{bmatrix} \begin{bmatrix} 3.5744 \\ 4.4140 \end{bmatrix}$	$\begin{bmatrix} 6.6475 \\ 17.5217 \\ 2.9930 \\ 0.7515 \end{bmatrix} \begin{bmatrix} 7.7984 \\ 2.9576 \\ 1.1647 \\ 7.7003 \end{bmatrix} \begin{bmatrix} 8.7400 \\ 3.3670 \\ 6.1257 \\ 2.4325 \end{bmatrix}$ $\begin{bmatrix} 0.9872 \\ 1.0732 \\ 3.0378 \\ 0.3062 \end{bmatrix} \begin{bmatrix} 3.9120 \\ 0.4951 \\ 4.5588 \\ 1.2267 \end{bmatrix} \begin{bmatrix} 10.2398 \\ 0.6033 \\ 2.4180 \\ 0.6774 \end{bmatrix}$

	X	F (Step 1, Eq. 3.2)
$l = 3$	$\begin{bmatrix} -0.4357 & -0.7494 & -0.3565 & -1.0186 & 1.0491 & 1.1541 \\ -0.4357 & -0.7494 & -0.3565 & -1.0186 & 1.0491 & 1.1541 \\ -0.9715 & 0.1627 & 0.0057 & 0.6932 & -1.6579 & -0.5544 \end{bmatrix}$	$\begin{bmatrix} 0.7515 & 1.1647 & 2.4325 \\ 0.7515 & 1.1647 & 2.4325 \\ 0.9703 & 0.4805 & 3.0562 \end{bmatrix}$

(continued)

Table 3.1 (continued)

	F^T (Step 4, Eq. 3.7)	F^{Tz} (Step 6, Eq. 3.10)
	$\begin{bmatrix} 0.3276 \\ 0.6218 \end{bmatrix} \begin{bmatrix} 1.7695 \\ 1.0992 \end{bmatrix} \begin{bmatrix} 0.0881 \\ 8.6904 \end{bmatrix}$ $\begin{bmatrix} 1.2772 \\ 6.1061 \end{bmatrix} \begin{bmatrix} 1.6018 \\ 2.0007 \end{bmatrix} \begin{bmatrix} 2.5886 \\ 11.1223 \end{bmatrix}$	$\begin{bmatrix} 10.4735 \\ 6.1670 \\ 1.6276 \\ 0.7038 \end{bmatrix} \begin{bmatrix} 3.1505 \\ 0.6858 \\ 1.5501 \\ 0.0470 \end{bmatrix} \begin{bmatrix} 0.9388 \\ 2.2083 \\ 6.8189 \\ 5.8672 \end{bmatrix}$ $\begin{bmatrix} 4.3181 \\ 2.0717 \\ 5.4549 \\ 1.4087 \end{bmatrix} \begin{bmatrix} 4.6674 \\ 0.4319 \\ 6.6115 \\ 1.2808 \end{bmatrix} \begin{bmatrix} 0.1611 \\ 7.0118 \\ 2.3274 \\ 0.3228 \end{bmatrix}$
	$\vdots$	$\vdots$
	$\vdots$	$\vdots$
	$\vdots$	$\vdots$
$l = 48$	X	F (Step 1, Eq. 3.2)
	$\begin{bmatrix} 4.98E-8 & 1.03E-7 & -2.95E-9 & -7.71E-8 & -8.29E-8 & -1.11E-7 \\ -4.33E-8 & 4.71E-8 & -2.95E-9 & -7.71E-8 & -8.29E-8 & -1.11E-7 \\ -2.59E-8 & 6.17E-8 & 6.5E-8 & 8.76E-8 & 1.03E-7 & -5.9E-9 \end{bmatrix}$	$\begin{bmatrix} 1.31E-14 & 5.96E-15 & 1.90E-14 \\ 4.09E-15 & 5.96E-15 & 1.90E-14 \\ 4.48E-15 & 1.12E-14 & 1.06E-14 \end{bmatrix}$
	F^T (Step 4, Eq. 3.7)	F^{Tz} (Step 6, Eq. 3.10)

(continued)

Table 3.1 (continued)

	$\begin{bmatrix} 2.05E-14 \\ 5.64E-14 \end{bmatrix} \begin{bmatrix} 3.59E-14 \\ 5.99E-14 \end{bmatrix} \begin{bmatrix} 9.92E-15 \\ 5.32E-14 \end{bmatrix}$ $\begin{bmatrix} 4.13E-14 \\ 2.92E-15 \end{bmatrix} \begin{bmatrix} 3.54E-14 \\ 9.02E-15 \end{bmatrix} \begin{bmatrix} 2.12E-14 \\ 3.58E-14 \end{bmatrix}$	$\begin{bmatrix} 1.80E-14 \\ 5.62E-15 \\ 3.76E-14 \\ 1.57E-14 \end{bmatrix} \begin{bmatrix} 2.54E-14 \\ 2.00E-14 \\ 1.95E-15 \\ 7.82E-14 \end{bmatrix} \begin{bmatrix} 1.05E-13 \\ 4.06E-14 \\ 2.19E-14 \\ 1.70E-14 \end{bmatrix}$ $\begin{bmatrix} 2.49E-15 \\ 2.38E-14 \\ 1.67E-15 \\ 2.30E-14 \end{bmatrix} \begin{bmatrix} 3.56E-15 \\ 1.35E-14 \\ 7.74E-15 \\ 2.36E-14 \end{bmatrix} \begin{bmatrix} 3.86E-14 \\ 6.24E-15 \\ 5.02E-14 \\ 3.03E-14 \end{bmatrix}$
$l = 49$	X	F(Step 1, Eq. 3.2)
	$\begin{bmatrix} 5.05E-8 & -5.54E-8 & 1.92E-8 & -3.98E-8 & -9.49E-8 & 3E-8 \\ 1.28E-9 & -5.4E-8 & -1.92E-8 & -3.98E-8 & -1.28E-7 & 2.46E-8 \\ -4.33E-8 & 4.71E-8 & 2.95E-9 & -7.77E-8 & 1.03E-7 & -5.9E-9 \end{bmatrix}$	$\begin{bmatrix} 5.62E-15 & 1.95E-15 & 9.92E-15 \\ 2.92E-15 & 1.95E-15 & 1.72E-15 \\ 4.09E-15 & 5.96E-14 & 1.06E-14 \end{bmatrix}$
	F^{T} (Step 4, Eq. 3.7)	F^{Tz} (Step 6, Eq. 3.10)
	$\begin{bmatrix} 3.34E-14 \\ 4.73E-15 \end{bmatrix} \begin{bmatrix} 7.96E-15 \\ 2.11E-14 \end{bmatrix} \begin{bmatrix} 1.78E-14 \\ 4.17E-14 \end{bmatrix}$ $\begin{bmatrix} 2.47E-14 \\ 7.68E-15 \end{bmatrix} \begin{bmatrix} 1.19E-14 \\ 8.36E-16 \end{bmatrix} \begin{bmatrix} 6.41E-15 \\ 1.33E-14 \end{bmatrix}$	$\begin{bmatrix} 2.38E-15 \\ 2.25E-14 \\ 1.86E-14 \\ 2.87E-14 \end{bmatrix} \begin{bmatrix} 4.28E-14 \\ 1.15E-14 \\ 1.32E-14 \\ 4.69E-14 \end{bmatrix} \begin{bmatrix} 2.79E-14 \\ 3.41E-14 \\ 1.09E-14 \\ 8.24E-16 \end{bmatrix}$ $\begin{bmatrix} 2.65E-14 \\ 3.08E-15 \\ 2.02E-14 \\ 3.53E-14 \end{bmatrix} \begin{bmatrix} 4.21E-14 \\ 3.54E-14 \\ 7.60E-15 \\ 3.37E-14 \end{bmatrix} \begin{bmatrix} 1.24E-14 \\ 3.10E-15 \\ 3.37E-14 \\ 6.32E-16 \end{bmatrix}$

(continued)

Table 3.1 (continued)

$l = 50$	X		F (Step 1, Eq. 3.2)
	$\begin{bmatrix} 3E-8 & -3.84E-8 & 8.31E-8 & 3.18E-8 & -8.6E-9\ 2.9E-8 \\ 3E-8 & -3.84E-8 & -2.78E-8 & 8E-9 & -8.6E-9\ 2.9E-8 \\ 1.28E-9 & 5.4E-8 & 1.92E-8 & -3.98E-8 & 9.4E-8\quad 3E-8 \end{bmatrix}$		$\begin{bmatrix} 2.4-15 & 7.96E-15 & 8.24E-16 \\ 2.4E-15 & 8.36E-16 & 8.24E-15 \\ 2.9E-15 & 1.95E-15 & 9.92E-15 \end{bmatrix}$

F^T (Step 4, Eq. 3.7)			F^{Tz} (Step 6, Eq. 3.10)
$\begin{bmatrix} 2.67E-15 \\ 9.29E-15 \end{bmatrix}\begin{bmatrix} 2.77E-15 \\ 1.15E-14 \end{bmatrix}\begin{bmatrix} 3.82E-16 \\ 1.29E-14 \end{bmatrix}$			$\begin{bmatrix} 7.15E-15 \\ 6.48E-15 \\ 6.32E-15 \\ 1.33E-14 \end{bmatrix}\begin{bmatrix} 1.11E-15 \\ 5.92E-15 \\ 7.86E-15 \\ 1.61E-14 \end{bmatrix}\begin{bmatrix} 9.89E-15 \\ 9.98E-15 \\ 1.32E-14 \\ 3.60E-14 \end{bmatrix}$
$\begin{bmatrix} 4.01E-15 \\ 8.98E-15 \end{bmatrix}\begin{bmatrix} 1.24E-14 \\ 1.15E-14 \end{bmatrix}\begin{bmatrix} 8.41E-15 \\ 8.02E-15 \end{bmatrix}$			$\begin{bmatrix} 2.93E-15 \\ 1.65E-14 \\ 1.64E-15 \\ 5.38E-15 \end{bmatrix}\begin{bmatrix} 2.05E-15 \\ 4.46E-15 \\ 1.69E-15 \\ 1.11E-14 \end{bmatrix}\begin{bmatrix} 2.44E-15 \\ 1.51E-14 \\ 2.53E-15 \\ 4.77E-15 \end{bmatrix}$

Learning attempt (l)	F^Z (Step 2, Eq. 3.3)	
$l = 2$	$\begin{bmatrix} 0.9703 & 0.4805 & 3.0562 \end{bmatrix}$	
	F (Step 8, Eq. 3.12)	Minimum
	$\begin{bmatrix} 0.7515 & 1.1647 & 2.4325 \\ 0.3062 & 0.4951 & 0.6033 \\ 0.9703 & 0.4805 & 3.0562 \end{bmatrix}$	0.3062

(continued)

Table 3.1 (continued)

$l = 3$	F^{Z} (Step 2, Eq. 3.3)	
	$\begin{bmatrix} 0.7515 & 0.4805 & 2.4325 \end{bmatrix}$	
	F (Step 8, Eq. 3.12)	Minimum
	$\begin{bmatrix} 0.3276 & 0.0470 & 0.0881 \\ 1.2772 & 0.4319 & 0.1611 \\ 0.7515 & 0.4805 & 2.4325 \end{bmatrix}$	0.0470
$\vdots$	$\vdots$	$\vdots$
$\vdots$	$\vdots$	$\vdots$
$\vdots$	$\vdots$	$\vdots$
$l = 48$	F^{Z} (Step 2, Eq. 3.3)	
	$\begin{bmatrix} 4.09E-15 & 5.96E-15 & 1.06E-14 \end{bmatrix}$	
	F (Step 8, Eq. 3.12)	Minimum
	$\begin{bmatrix} 5.62E-15 & 1.95E-15 & 9.92E-15 \\ 1.67E-15 & 3.56E-15 & 6.24E-15 \\ 4.09E-15 & 5.96E-15 & 1.06E-14 \end{bmatrix}$	1.67E-15

(continued)

Table 3.1 (continued)

$l = 49$	F^Z (Step 2, Eq. 3.3)	
	$\begin{bmatrix} 2.92E-15 & 1.95E-15 & 1.72E-15 \end{bmatrix}$	
	F (Step 8, Eq. 3.12)	Minimum
	$\begin{bmatrix} 2.38E-15 & 7.96E-15 & 8.24E-16 \\ 3.08E-15 & 8.36E-16 & 3.10E-15 \\ 2.92E-15 & 1.95E-15 & 1.72E-15 \end{bmatrix}$	8.24E-16
$l = 50$	F^Z (Step 2, Eq. 3.3)	
	$\begin{bmatrix} 2.4E-15 & 8.36E-16 & 8.24E-16 \end{bmatrix}$	
	F (Step 8, Eq. 3.12)	Final solution f^*
	$\begin{bmatrix} 2.67E-15 & 1.11E-15 & 3.82E-16 \\ 1.64E-15 & 1.69E-15 & 2.44E-15 \\ 2.4E-15 & 8.36E-15 & 8.24E-16 \end{bmatrix}$	3.82E-16

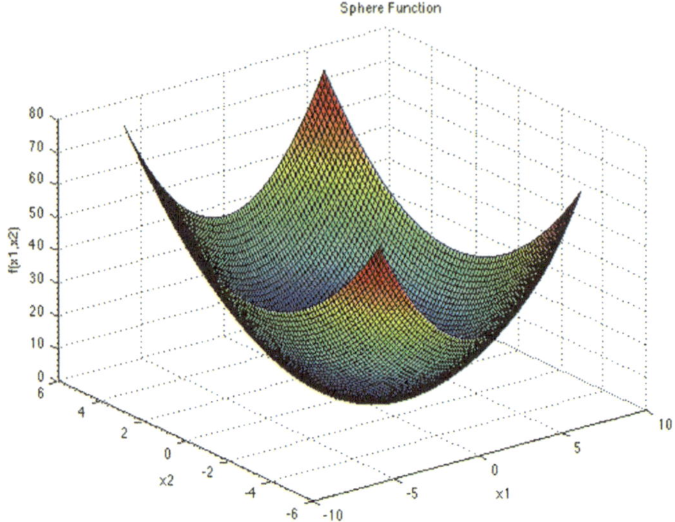

Fig. 3.2　3D plot for Sphere function

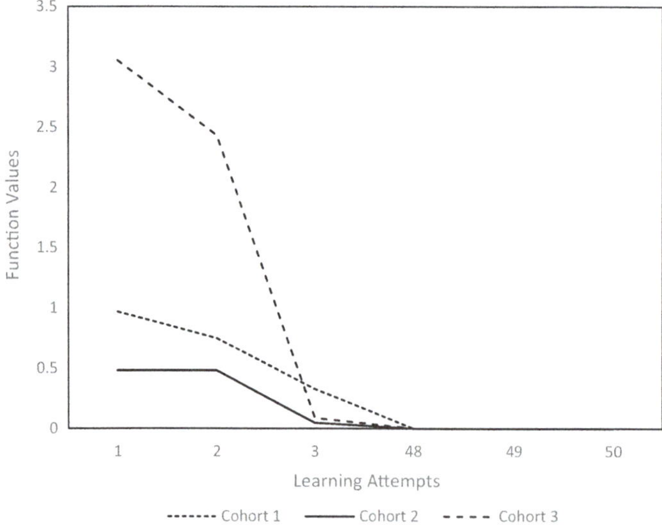

Fig. 3.3　Convergence for Sphere function

(Step 3, Eq. 3.4):

$$p_1^1 = 0.7476$$
$$p_1^2 = 0.2524$$

$$p_2^1 = 0.4943$$

$$p_2^3 = 0.5057$$

$$p_3^1 = 0.2876$$

$$p_3^2 = 0.7124$$

(Step 4, Eq. 3.5**):**

Now every cohort $k(k = 1, \ldots, K)$ is left with $C^k - 1$ candidates. Consider candidate C_1 in cohort 1 and the associated qualities $X_1^{C_1} = \left[x_{1,1}^1, x_{2,1}^1 \right] = [0.4426, 2.7681]$. Sampling interval for $x_{1,1}^1$ is given by

$$\left[\psi_1^{1,lower}, \psi_1^{1,upper}\right] = \begin{bmatrix} 0.4426 - \left(\dfrac{5.12 - (-5.12)}{2} \right) \times 0.98, \\ 0.4426 + \left(\dfrac{5.12 - (-5.12)}{2} \right) \times 0.98 \end{bmatrix}$$

$$[\psi_1^{1,lower}, \psi_1^{1,upper}] = [-4.575, 5.4602]$$

$= [-4.575, 5.12](\because 5.4602 \text{ is out of the interval})$

Sampling interval for $x_{2,1}^1$ is given by

$$[\psi_2^{1,lower}, \psi_2^{1,upper}] = \begin{bmatrix} 2.7681 - \left(\dfrac{5.12 - (-5.12)}{2} \right) \times 0.98, \\ 2.7681 + \left(\dfrac{5.12 - (-5.12)}{2} \right) \times 0.98 \end{bmatrix}$$

$$[\psi_2^{1,lower}, \psi_2^{1,upper}] = [7.7807, 2.2545]$$

$= [5.12, 2.2545](\because -7.7807 \text{ is out of the interval})$

Using roulette wheel selection, candidate C_2 in cohort 1 chooses to follow candidate C_1. So sampling intervals for candidate C_2 will be the same as that of candidate C_1. The $T = 2$ sampling intervals of every candidate $C^k - 1, k(k = 1, \ldots, K)$ are as follows:

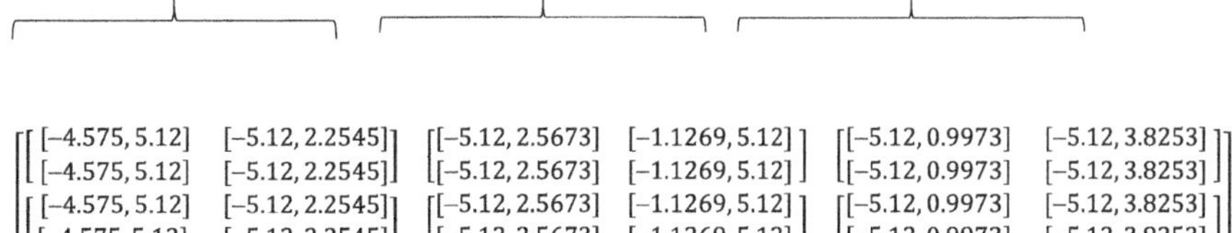

$$\begin{bmatrix} \begin{bmatrix} [-4.575, 5.12] & [-5.12, 2.2545] \\ [-4.575, 5.12] & [-5.12, 2.2545] \end{bmatrix} & \begin{bmatrix} [-5.12, 2.5673] & [-1.1269, 5.12] \\ [-5.12, 2.5673] & [-1.1269, 5.12] \end{bmatrix} & \begin{bmatrix} [-5.12, 0.9973] & [-5.12, 3.8253] \\ [-5.12, 0.9973] & [-5.12, 3.8253] \end{bmatrix} \\ \begin{bmatrix} [-4.575, 5.12] & [-5.12, 2.2545] \\ [-4.575, 5.12] & [-5.12, 2.2545] \end{bmatrix} & \begin{bmatrix} [-5.12, 2.5673] & [-1.1269, 5.12] \\ [-5.12, 2.5673] & [-1.1269, 5.12] \end{bmatrix} & \begin{bmatrix} [-5.12, 0.9973] & [-5.12, 3.8253] \\ [-5.12, 0.9973] & [-5.12, 3.8253] \end{bmatrix} \end{bmatrix}$$

(Step 4, Eq. 3.6**):** Every candidate samples the qualities from within these sampling intervals and forms quality matrix $\mathbf{Z}^T$ as follows:

$$
\mathbf{Z}^T = \begin{bmatrix}
\begin{bmatrix} 3.7775 & -4.9854 \\ -0.0271 & 2.1268 \end{bmatrix} &
\begin{bmatrix} -2.4902 & 4.1556 \\ -2.8815 & 4.4581 \end{bmatrix} &
\begin{bmatrix} -1.6575 & -0.5544 \\ -1.3713 & -4.2098 \end{bmatrix} \\[2ex]
\begin{bmatrix} -1.5062 & -0.00348 \\ -2.2765 & -2.2535 \end{bmatrix} &
\begin{bmatrix} -4.7380 & 4.1285 \\ -1.7774 & 2.8955 \end{bmatrix} &
\begin{bmatrix} -4.4906 & -2.1964 \\ -4.65257 & -1.0430 \end{bmatrix}
\end{bmatrix}
$$

(**Step 4, Eq.** 3.7):
$$
\mathbf{F}^T = \begin{bmatrix}
\begin{bmatrix} 39.1242 \\ 4.5240 \end{bmatrix} &
\begin{bmatrix} 23.4702 \\ 28.1783 \end{bmatrix} &
\begin{bmatrix} 3.0562 \\ 19.6036 \end{bmatrix} \\[2ex]
\begin{bmatrix} 2.2700 \\ 11.6100 \end{bmatrix} &
\begin{bmatrix} 39.4941 \\ 11.5436 \end{bmatrix} &
\begin{bmatrix} 24.9901 \\ 22.4855 \end{bmatrix}
\end{bmatrix}
$$

(**Step 5, Eq.** 3.8):

$$
p_1^1 = 0.3065
$$
$$
p_2^2 = 0.1041
$$
$$
p_3^3 = 0.5894
$$

(**Step 6, Eq.** 3.9):

$$
\mathbf{Z}^{T_p} = \begin{bmatrix}
\begin{bmatrix} -0.9714 & 0.1627 \\ -4.5412 & -2.4676 \\ -4.0294 & -0.5608 \\ -4.2622 & -1.8622 \end{bmatrix} &
\begin{bmatrix} 0.0057 & 0.6931 \\ -4.8913 & 2.3370 \\ -4.4565 & -1.9554 \\ -1.5795 & 1.0522 \end{bmatrix} &
\begin{bmatrix} -2.9102 & -3.4787 \\ -2.9835 & -0.8647 \\ -2.1960 & -4.7693 \\ -2.2042 & -3.3672 \end{bmatrix} \\[4ex]
\begin{bmatrix} 1.1820 & -2.2922 \\ -0.9737 & -3.0238 \\ -1.7371 & -1.50645 \\ 0.0012 & 2.3825 \end{bmatrix} &
\begin{bmatrix} 1.2259 & -3.5126 \\ -0.6327 & -0.3592 \\ -0.2333 & -2.3488 \\ -2.5639 & -1.2459 \end{bmatrix} &
\begin{bmatrix} 0.7412 & -2.7817 \\ -3.0381 & -0.8921 \\ -0.3807 & 2.1570 \\ -2.7075 & -0.9127 \end{bmatrix}
\end{bmatrix}
$$

(**Step 6, Eq.** 3.10):

$$
\mathbf{F}^{T_p} = \begin{bmatrix}
\begin{bmatrix} 0.9703 \\ 26.7122 \\ 16.5508 \\ 21.6350 \end{bmatrix} &
\begin{bmatrix} 0.4805 \\ 29.387 \\ 23.6842 \\ 3.6022 \end{bmatrix} &
\begin{bmatrix} 20.5714 \\ 9.6495 \\ 27.5692 \\ 18.3430 \end{bmatrix} \\[4ex]
\begin{bmatrix} 9.9358 \\ 10.0916 \\ 5.2871 \\ 5.6763 \end{bmatrix} &
\begin{bmatrix} 13.8416 \\ 0.5294 \\ 5.5713 \\ 8.1260 \end{bmatrix} &
\begin{bmatrix} 8.2876 \\ 10.0264 \\ 4.7976 \\ 8.1636 \end{bmatrix}
\end{bmatrix}
$$

(Step 8, Eq. 3.12):

$$F = \begin{bmatrix} 0.9703 & 0.4805 & 3.0562 \\ 2.2700 & 0.5294 & 4.7976 \\ 6.8402 & 20.1344 & 3.5564 \end{bmatrix}$$

$$Minimum = 0.4805$$

End of Learning Attempt

3.3 Result Analysis of Multi-CI

The Multi-CI algorithm was coded in MATLAB R2013a on Windows Platform with a T6400@4 GHz Intel Core 2 Duo processor with 4 GB RAM. Executable code is given on the following website: https://sites.google.com/site/oatresearch.

The algorithm was validated in [14] by solving two well studied sets of test problems. Set 1 included 50 well studied benchmark problems [4, 5]. Set 2 included 25 test problems from CEC 2005 [17]. In this chapter few Set 1 and Set 2 test problems are solved which are listed in Tables 3.2 and 3.3 respectively. Every problem in these

Table 3.2 The benchmark problems used in Test 1

Problem	Name	Type	Low	Up	Dimensions
F1	Foxholes	MS	−65.536	65.536	2
F5	Ackley	MN	−32	32	30
F6	Beale	UN	−4.5	4.5	5
F16	Fletcher	MN	−3.1416	3.1416	5
F28	Michalewics10	MS	0	3.1416	10
F43	Six-hump camelback	MN	−5	5	2

Dim Dimension; *Low and Up* Limitations of search space; *U* Unimodal; *M* Multimodal; *S* Separable; *N* Non-separable

Table 3.3 The benchmark problems used in Test 2

Problem	Name	Type	Low	Up	Dimensions
F51	Shifted sphere	U	−100	100	10
F56	Shifted Rosenbrock's	M	−100	100	10
F63	Expanded extended Griewank's + Rosenbrock's	E	−3	1	10
F75	Rotated hybrid comp. Fn 4	HC	−2	5	10

Dim Dimension; *Low and Up* Limitations of search space; *U* Unimodal; *M* Multimodal; *E* Expanded; *H* Hybrid)

test cases was solved 30 times using Multi-CI. In every run, initial behaviour of every candidate was randomly initialized.

3.3.1   Control Parameters and Stopping Criteria

The Multi-CI parameters chosen were as follows:

- Number of cohorts $K = 3$
- Number of candidates $C^k = 5$
- Reduction factor value $r = 0.98$
- Quality variation parameters $T = 5$ and $T_p = 10$

  The algorithm stopped when any of the following condition is satisfied.

- Objective function value is less than 10^{-16}
- Maximum number of function evaluations 20,000
- Maximum number of learning attempts reached.

3.3.2   Statistical Analysis

The Multi-CI algorithm presented here and the other algorithms with which the results are being compared are stochastic in nature due to which in every independent run of the algorithm the converged solution may be different than one another. A pairwise comparison of Multi-CI and every other algorithm was carried out, i.e. the converged (global minimum) values of 30 independent runs solving every problem using Multi-CI were compared with every other algorithm solving 30 independent runs of these problems. The Wilcoxon Signed-Rank test was used for such pairwise comparison. Similar to [1], the significance value α was chosen to be 0.05 with null hypothesis H0 is: There is no difference between the median of the solutions obtained by algorithm A and the median of the solutions obtained by algorithm B for the same set of test problems, i.e. median (A) = median(B). Also, to determine whether algorithm A yielded statistically better solution than algorithm B or whether alternative hypothesis was valid, the sizes of the ranks provided by the Wilcoxon Signed-Rank test (T + and T-) were thoroughly examined.

The mean solution, best solution and standard deviation (SD.), mean run time (in seconds) over the 30 runs of the algorithms solving few Test 1 and Test 2 problems are represented in Table 3.4 and Table 3.5, respectively. The multi problem based pair wise analysis using the averages of the global solutions obtained over the 30 runs of the algorithms solving the Test 1 and Test 2 problems are presented in Table 3.6. The results highlighted that the Multi-CI algorithm performed significantly better than every other algorithm.

The convergence plots of few unimodal and multimodal representative functions such as Ackley function, Beale function, Fletcher function, Foxhole functions,

Table 3.4 Statistical solutions to Test 1 problems using PSO, CMAES, ABC, CLPSO, SADE, BSA, IA and Multi-CI

Problem	Statistics	PSO2011	CMAES	ABC	JDE	CLPSO	SADE
F1	Mean	1.3316029264876300	10.0748846367972000	0.9980038377944500	1.0641405484285200	1.8209961275956800	0.9980038377944500
	SD	0.9455237994690700	8.0277365400340800	0.0000000000000001	0.3622456829347420	1.6979175079427900	0.0000000000000000
	Best	0.9980038377944500	0.9980038377944500	0.9980038377944500	0.9980038377944500	0.9980038377944500	0.9980038377944500
	Runtime	72.527	44.788	64.976	51.101	61.650	66.633
F5	Mean	1.5214322973725000	11.7040011684582000	0.0000000000000340	0.0811017056422860	0.1863456353861950	0.7915368220335460
	SD	0.6617570384662600	9.7201961540865200	0.0000000000000035	0.3176012689149320	0.4389839299322230	0.7561593402959740
	Best	0.0000000000000080	0.0000000000000080	0.0000000000000293	0.0000000000000044	0.0000000000000080	0.0000000000000044
	Runtime	63.039	3.144	23.293	11.016	45.734	40.914
F6	Mean	0.0000000041922968	0.2540232169641050	0.0000000000000028	0.0000000000000000	0.0000444354499943	0.0000000000000000
	SD	0.0000000139615552	0.3653844307786430	0.0000000000000030	0.0000000000000000	0.0001015919507724	0.0000000000000000
	Best	0.0000000000000000	0.0000000000000000	0.0000000000000005	0.0000000000000000	0.0000000000000000	0.0000000000000000
	Runtime	32.409	4.455	22.367	1.279	125.839	4.544
F16	Mean	48.7465164446927000	1680.3460230073400000	0.0218688498331872	0.9443728655432830	81.7751618148164000	0.0000000000000000
	SD	88.8658510972991000	2447.7484859066000000	0.0418409568792831	2.8815514827061600	379.9241117377270000	0.0000000000000000
	Best	0.0000000000000000	0.0000000000000000	0.0000000000000016	0.0000000000000000	0.0000000000000000	0.0000000000000000
	Runtime	95.352	11.947	44.572	4.719	162.941	5.763
F28	Mean	−8.9717330307549300	−7.6193507368464700	−9.6601517156413500	−9.6397230986132500	−9.6400278592589600	−9.6572038232921700
	SD	0.4927013165009220	0.7904830398850970	0.0000000000000008	0.0393668145094111	0.0437935551332868	0.0105890022905617
	Best	−9.5777818097208200	−9.1383975057875100	−9.6601517156413500	−9.6601517156413500	−9.6601517156413500	−9.6601517156413500
	Runtime	144.093	6.959	27.051	20.803	32.801	46.395
F43	Mean	−1.0316284534898800	−1.0044229658530100	−1.0316284534898800	−1.0316284534898800	−1.0316284534898800	−1.0316284534898800

(continued)

Table 3.4 (continued)

Problem	Statistics	PSO2011	CMAES	ABC	JDE	CLPSO	SADE
	SD	0.0000000000000005	0.1490105926664260	0.0000000000000005	0.0000000000000005	0.0000000000000005	0.0000000000000005
	Best	−1.0316284534898800	−1.0316284534898800	−1.0316284534898800	−1.0316284534898800	−1.0316284534898800	−1.0316284534898800
	Runtime	16.754	24.798	11.309	7.147	18.564	27.650

Problem	Statistics	BSA	IA	Multi CI
F1	Mean	0.9980038377944500	0.9980038690000000	0.9980038377944500
	SD	0.0000000000000000	0.0000000000000035	0.0000000000000003
	Best	0.9980038377944500	0.9980038685998520	0.9980038377944500
	Runtime	38.125	43.535	1.092
F5	Mean	0.0000000000000105	0.0000000000000009	0.0000000000000000
	SD	0.0000000000000034	0.0000000000000000	0.0000000000000000
	Best	0.0000000000000080	0.0000000000000009	0.0000000000000000
	Runtime	14.396	49.458	5.243
F6	Mean	0.0000000000000000	0.0082236060000000	0.0000000000000000
	SD	0.0000000000000000	0.0000000000000000	0.0000000000000000
	Best	0.0000000000000000	0.0082236059357692	0.0000000000000000
	Runtime	0.962	50.246	1.356
F16	Mean	0.0000000000000000	0.0000000000000000	0.0000000000000000
	SD	0.0000000000000000	0.0000000000000000	0.0000000000000000
	Best	0.0000000000000000	0.0000000000000000	0.0000000000000000
	Runtime	7.781	48.262	0.459
F28	Mean	−9.6601517156413500	−6.2086254390000000	−8.4871985036037100

(continued)

Table 3.4 (continued)

Problem	Statistics	BSA	IA	Multi CI
	SD	0.0000000000000007	0.0000000000000027	0.2867921564163950
	Best	−9.6601517156413500	−6.2086254392105500	−8.9978275376597000
	Runtime	22.250	71.652	4.784
F43	Mean	−1.0316284534898800	−1.0304357800000000	−1.0316284534898800
	SD	0.0000000000000005	0.0014911900000000	0.0000000000000000
	Best	−1.0316284534898800	−1.0314500753985900	−1.0316284534898800
	Runtime	5.691	39.897	0.391

Mean Mean solution; *SD* Standard-deviation of mean solution; *Best* Best solution; *Runtime* Mean runtime in seconds

Table 3.5 Statistical solutions to Test 2 problems using PSO, CMAES, ABC, CLPSO, SADE, BSA, IA and Multi-CI

Problem	Statistics	PSO2011	CMAES	ABC	JDE	CLPSO	SADE
F51	Mean	−450.0000000000000000	−450.0000000000000000	−450.0000000000000000	−450.0000000000000000	−450.0000000000000000	−450.0000000000000000
	SD	0.0000000000000000	0.0000000000000000	0.0000000000000000	0.0000000000000000	0.0000000000000000	0.0000000000000000
	Best	−450.0000000000000000	−450.0000000000000000	−450.0000000000000000	−450.0000000000000000	−450.0000000000000000	−450.0000000000000000
	Runtime	212.862	23.146	113.623	118.477	167.675	154.232
F56	Mean	393.4959999056240000	390.5315438816460000	391.2531452421960000	231.3986579112350000	405.5233436479650000	390.2657719408230000
	SD	16.0224965900462000	1.3783433976378300	3.7254660805238600	247.2968415284400000	10.7480096852869000	1.0114275384776600
	Best	390.0000000000150000	390.0000000000000000	390.0101471658490000	−140.0000000000000000	390.5776683413440000	390.0000000000000000
	Runtime	1178.079	27.894	159.762	153.715	1441.859	1214.303
F63	Mean	−129.2373581503910000	−128.7850616923410000	−129.8343428775830000	−129.6294851450880000	−129.8382867796110000	−129.7129164862680000
	SD	0.5986210944493790	0.6157633658946230	0.0408016481905455	0.1054759371085400	0.0372256921835666	0.0875456568200232
	Best	−129.6861385930680000	−129.5105509483130000	−129.9098920058450000	−129.8125711770830000	−129.9098505660780000	−129.8717592632560000
	Runtime	2183.218	25.496	205.194	186.347	1526.365	660.986
F75	Mean	1107.9038127876700000	1401.6553278264300000	930.4565414149210000	1072.9924659809200000	1258.5157766524700000	1074.3695435628600000
	SD	127.9566489362040000	253.2428066220210000	87.9959072391079000	2.2606058314671500	241.4024507676890000	2.8314182838917800
	Best	1069.5511765775700000	1072.4973401423200000	862.4476004191700000	1068.5560012648600000	871.8607884176050000	1069.8723890709000000
	Runtime	4060.091	214.580	2113.339	2951.018	5262.210	3410.902

Problem	Statistics	BSA	IA	Multi CI
F51	Mean	−450.0000000000000000	−447.6018854297170000	−450.0000000000000000
	SD	0.0000000000000000	89.3142986500000000	0.0000000000000000
	Best	−450.0000000000000000	−450.0000000000000000	−450.0000000000000000
	Runtime	140.736	30.282	28.930

(continued)

Table 3.5 (continued)

Problem	Statistics	BSA	IA	Multi CI
F56	Mean	390.1328859704120000	390.8036739982730000	392.3754700583880000
	SD	0.7278464357038200	0.0000000000000000	0.6527183145462900
	Best	390.0000000000000000	390.8036739982730000	391.2787609196740000
	Runtime	290.236	45.632	88.645
F63	Mean	−129.8981409848090000	−122.2126680000000000	−129.3840699141840000
	SD	0.0682328484314248	0.0000000000000434	0.0756137409573255
	Best	−129.9901230990300000	−122.2126679617240000	−129.5432045245170000
	Runtime	1064.114	46.260	170.218
F75	Mean	1063.7363787709700000	471.2797518000000000	1084.7073068225200000
	SD	55.8479313799755000	2.2346287190000000	4.9504851126832800
	Best	856.8214538442850000	469.3372925643150000	1078.2231646698500000
	Runtime	4280.901	263.829	711.530

Mean Mean solution; *SD* Standard-deviation of mean solution; *Best* Best solution; *Runtime* Mean runtime in seconds

Table 3.6 Multi-problem based statistical pairwise comparison of PSO, CMAES, ABC, JDE, CLPSO, SADE, BSA, IA and Multi-CI

Other Algorithm versus Multi-CI	p-Value	$T+$	$T-$	Winner
PSO versus Multi-CI	0.0035	368	1010	Multi-CI
CMAES versus Multi-CI	3.3367e-07	235	1656	Multi-CI
ABC versus Multi-CI	0.1355	615	981	Multi-CI
JDE versus Multi-CI	4.6305e-04	320	1111	Multi-CI
CLPSO versus Multi-CI	1.507e-04	366	1345	Multi-CI
SADE versus Multi-CI	0.2031	424	657	Multi-CI
BSA versus Multi-CI	0.9144	402	418	Multi-CI
IA versus Multi-CI	0.7003	834	936	Multi-CI

Michalewics function, Six-hump camelback function are presented in Figs. 3.4, 3.5, 3.6, 3.7, 3.8 and 3.9. The best solutions in every learning attempt are also plotted in Figs. 3.4b, 3.5b, 3.6b, 3.7b, 3.8b and 3.9b. These plots exhibited the candidates' self-supervised intra as well as inter cohort learning behaviour. The convergence also highlighted the significance of Multi-CI approach quickly reaching the optimum solution.

This section provides theoretical comparison of the algorithms being compared with Multi-CI. The method of PSO a swarm of solutions modify their positions in the search space. Every particle of the swarm represents a solution which moves with certain velocity in the search space based on the best solution in the entire swarm as well as the best solution in certain close neighborhood. It imparts exploration as well as exploitation abilities to entire swarm. According to Teo et al. [18], Li and Yao [8] and Selvi and Umrani [13] the PSO may not be efficient solving the problems with discrete search space as well as non-coordinate systems and may need supporting techniques to solve such problems. In this chapter Multi-CI is compared with the advanced versions of the PSO referred to as Comprehensive Learning PSO (CLPSO) (Liang et al. 2016) and PSO2011 [11]. The technique of CMAES [3] is a mathematical-based algorithm which exploits adaptive mutation parameters through computing a covariance matrix. The computational cost of the covariance matrix calculation, sampling using multivariate normal distribution and factorization of covariance matrix may increase exponentially with increase in problem dimension [13].The algorithm of ABC [4] carries out exploration using random search by scout bees and exploitation using employed bees. Some studies highlightedthat the algorithm of ABC can perform well with exploration; however, it is not efficient in local search and exploitation [10]. This may make the algorithm trap into local minima. The BSA [1] is a populations based technique which deploys genetic operators to generate initial solutions. Then therandomly chooses individuals to find the new solutions in the search space. The non-uniform crossover makes BSA unique and powerful technique. Similar to the BSA, DE [12, 16] is also a population based technique which exploits genetic operators. The search process is mainly driven by the mutation and selection operation. The crossover operator is further deployed for

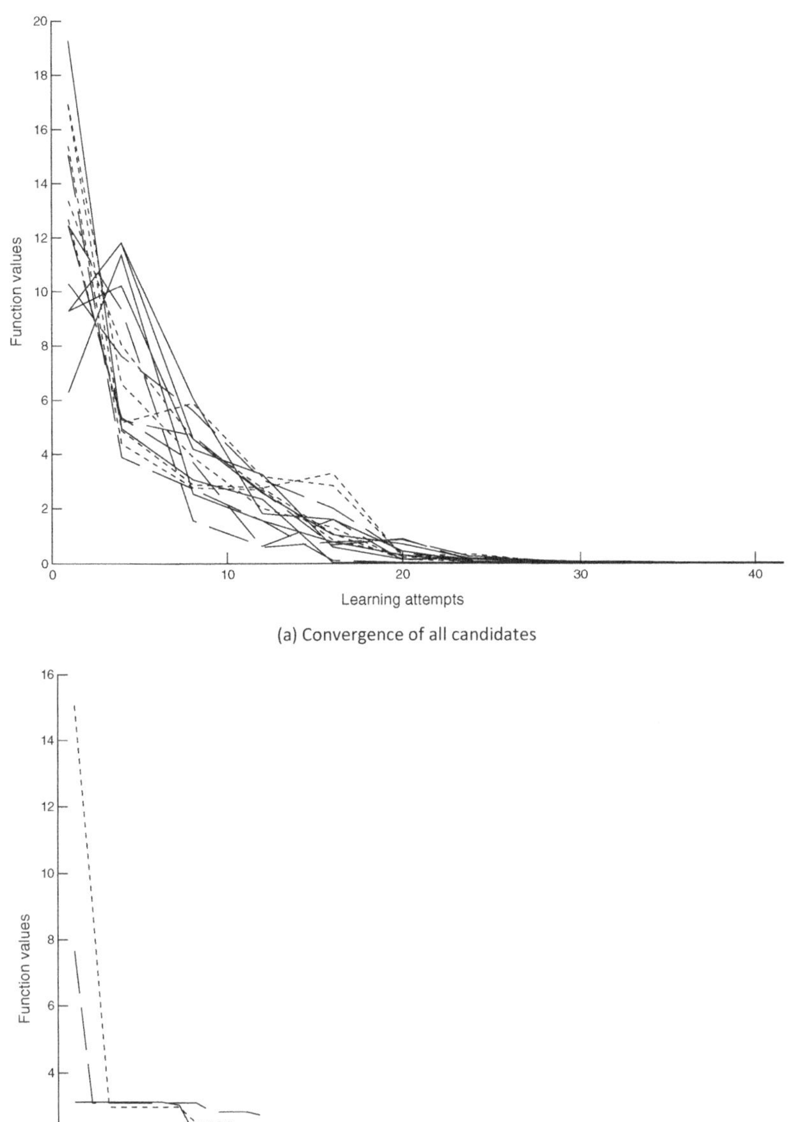

Fig. 3.4   Convergence for Ackley function (F5

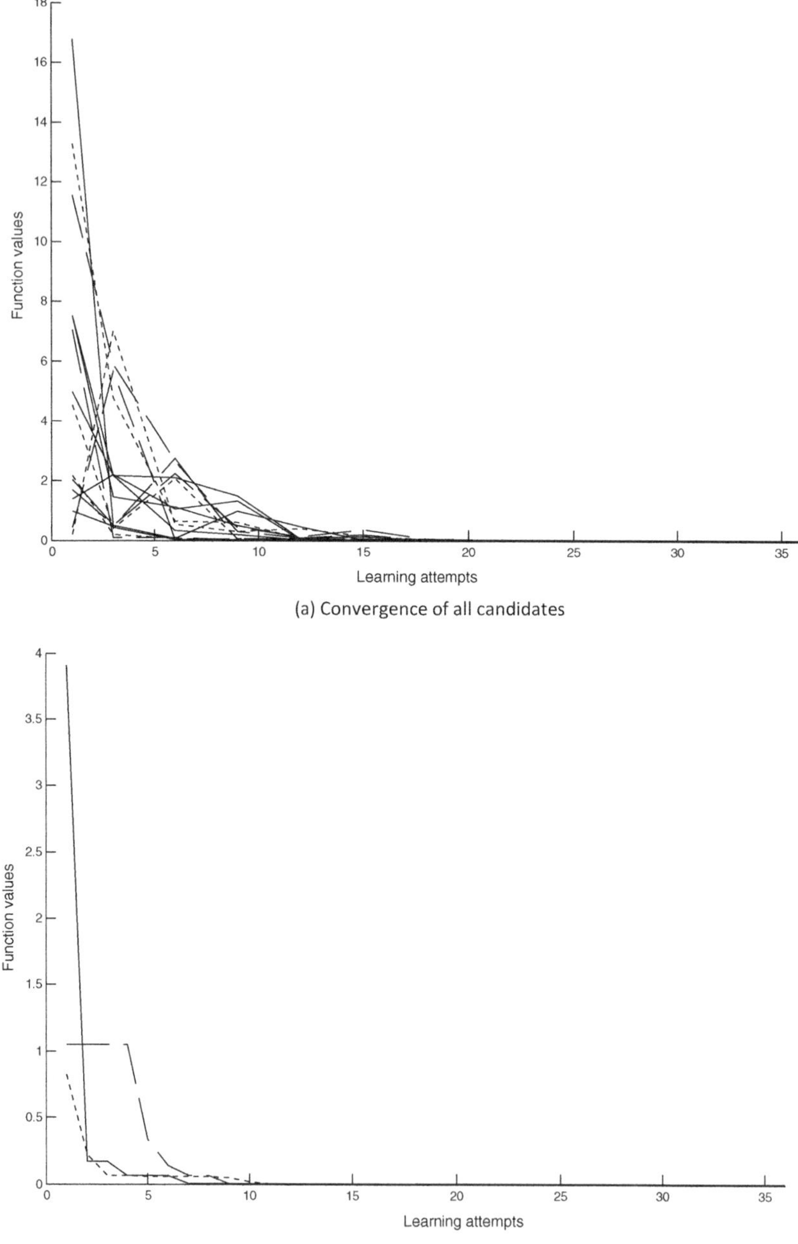

(a) Convergence of all candidates

(b) Convergence of best solutions

Fig. 3.5 Convergence for Beale function (F6)

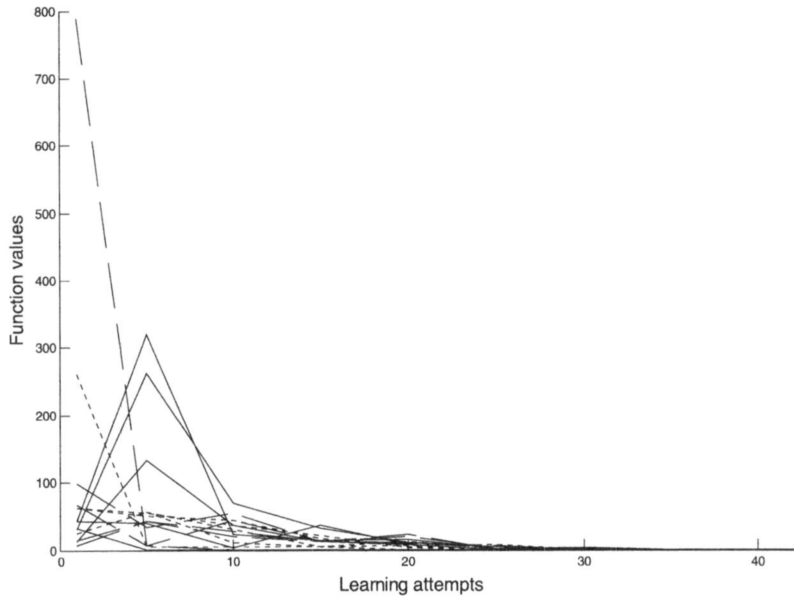

(a) Convergence of all candidates

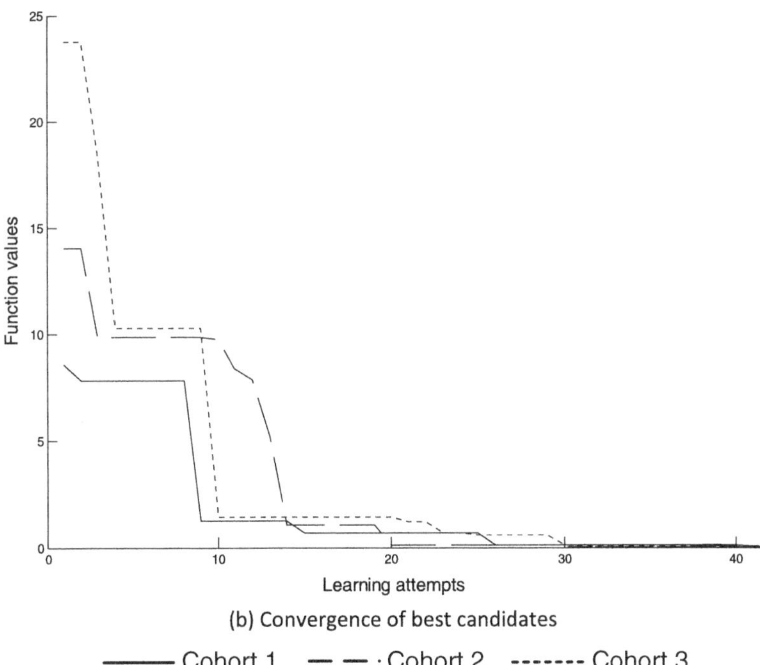

(b) Convergence of best candidates

Fig. 3.6 Convergence for Fletcher function (F16)

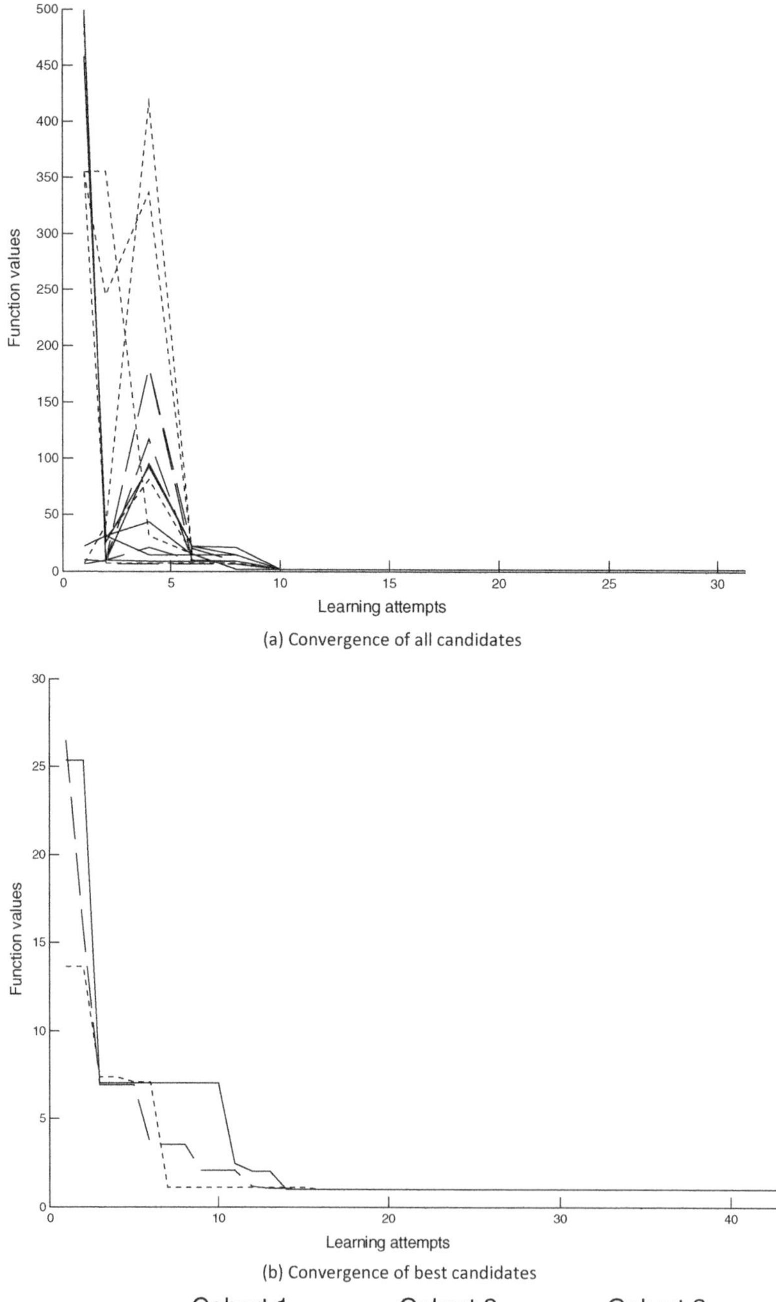

Fig. 3.7 Convergence for Foxholes function (F1)

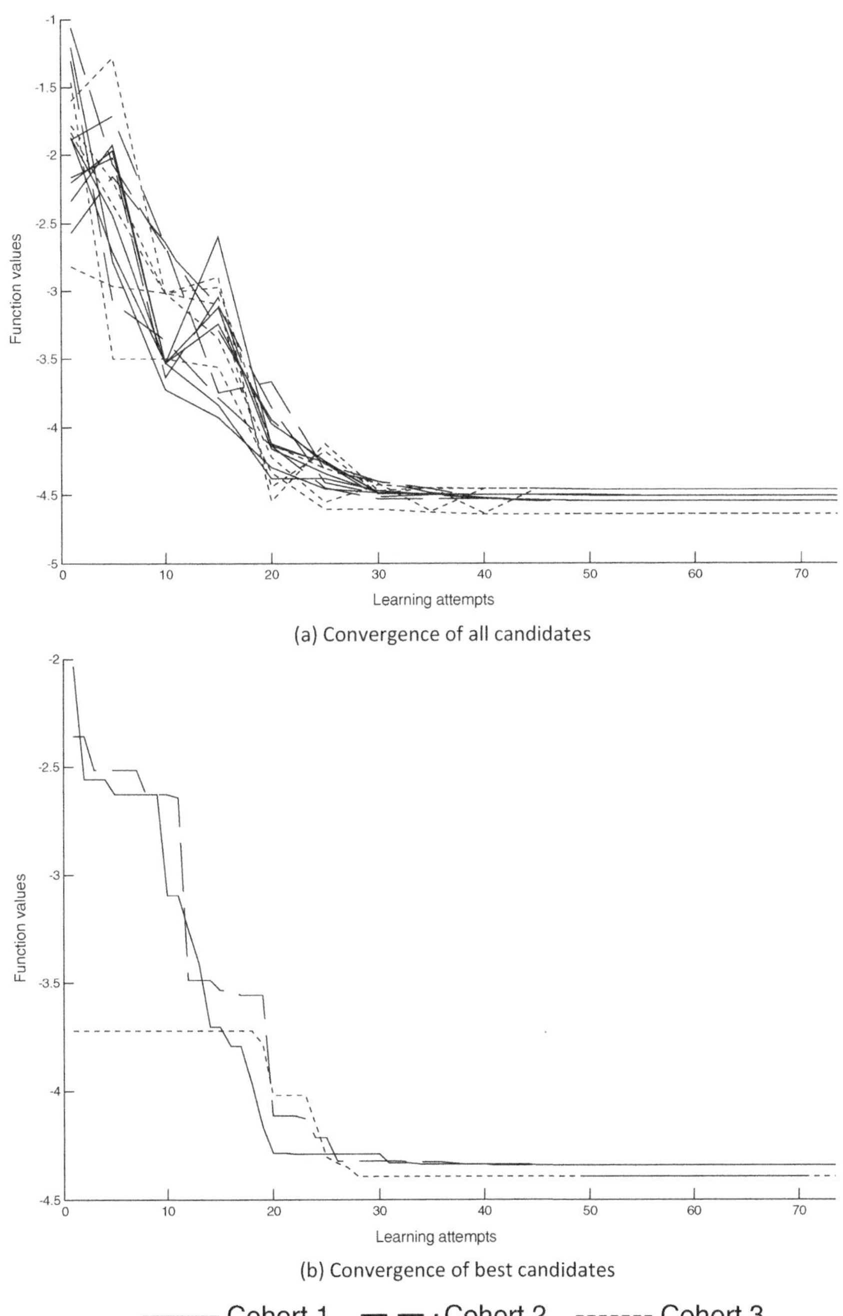

(a) Convergence of all candidates

(b) Convergence of best candidates

Fig. 3.8   Convergence for Michalewics function (F28)

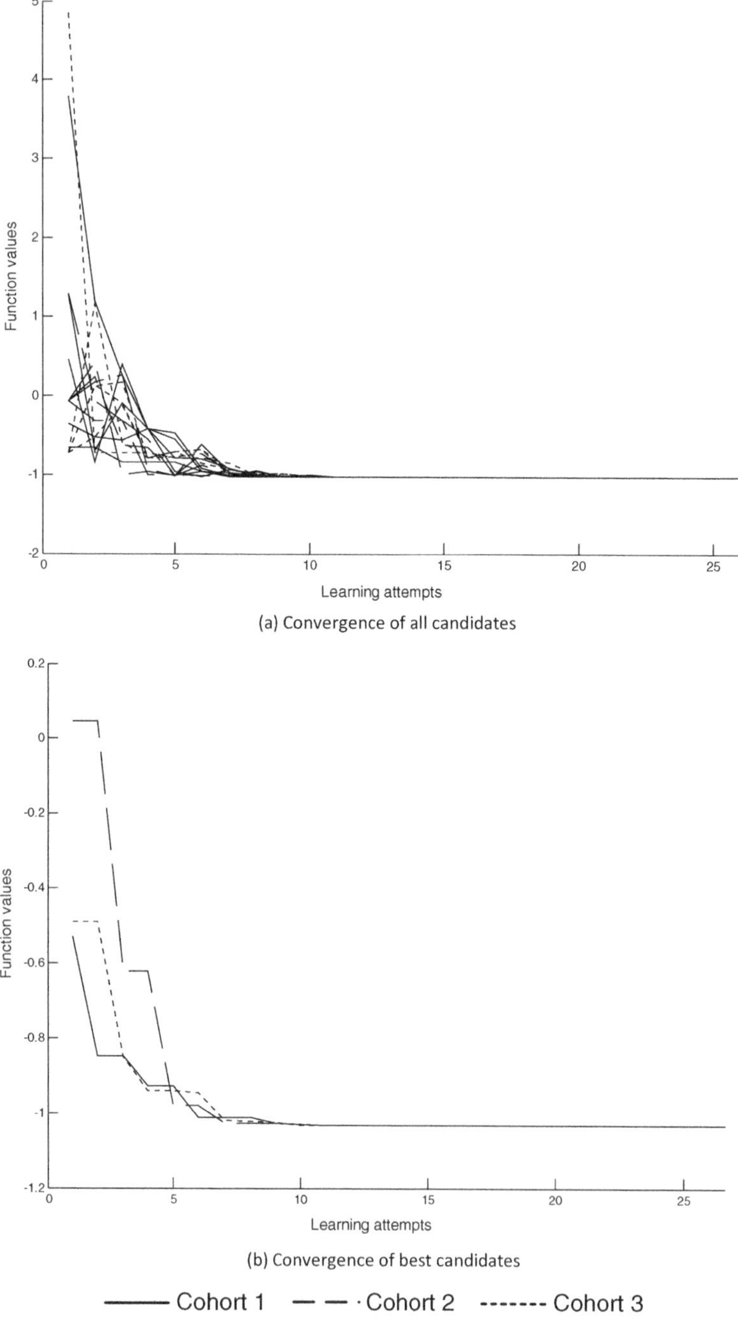

Fig. 3.9 Convergence for Six-hump camelback function (F43)

effectively sorting the trial vectors which helps to choose and retain better solutions. Teo et al. [18] recently proposed IA. It is inspired from competitive behavior of political party individuals. The local party leaders exploit the concepts such as introspection, local competition and global competition improving the solution quality through exploitation and exploration. In addition, the ordinary party members may follow the own party leader or other party leader. This changes the priority of the search lead by certain party. The algorithm performed better as compared to most of the contemporary algorithms.

The Multi-CI algorithm proposed here exhibited certain prominent characteristics and limitations. These are discussed below.

1. In Multi-CI the best individual from within every cohort are moved to a separate pool Z. One best candidate is chosen from within each cohort. Then every candidate chooses the best behavior/objective function value from within the $T + T_p$ choices. Thus every candidate competes with its own local best behavior as well as the best behavior chosen from the other cohorts. This gives more exploitation power to the algorithm due to which chance of avoiding the local minima increases with faster convergence.
2. The elite candidate behaviours from Pool Z is carry forwarded to the subsequent learning attempt. This helps not to lose the best behaviour (solution) so far and also the influence of such solutions does not diminish if not followed by any candidate.
3. Initial random walks of individuals around the cohorts ensure exploration of the search space around the candidates.
4. The Multi-CI parameters could be easily tuned which may make it a flexible algorithm for handling variety of problems with different dimensions and complexity.
5. The results highlighted that the algorithm is sufficiently robust with reasonable computational cost and is successful at exploring multi-modal search spaces.
6. The computational performance was essentially governed by sampling interval reduction factor r. Its value was chosen based on the preliminary trials of the algorithm.

3.4 Summary

A modified version of the Cohort Intelligence (CI) algorithm referred to as Multi-Cohort Intelligence (Multi-CI) was proposed. In the proposed Multi-CI approach intra-group learning and inter-group learning mechanisms were implemented. It is more realistic representation of the learning through interaction and competition of the cohort candidates. It imparted the exploitation and exploration capabilities to the algorithm. The approach was validated by solving two sets of test problems from CEC 2005. Wilcoxon statistical tests were conducted for comparing the performance of the algorithm with the existing algorithms. The performance of Multi-CI was exceedingly better as compared to PSO2011, CMAES, ABC, JDE, CLPSO and

SADE in terms of objective function value (best and mean), robustness, as well as computational time. The performance of the Multi-CI was marginally better as compared to BSA and IA. The solution quality highlighted that the Multi-CI is a robust approach with reasonable computational cost and could quickly reach in the close neighbourhood of the global optimum solution. In addition, Multi-CI algorithm has been applied on Advanced Machining Process problems by [15].

A generalized constraint handling mechanism needs to be developed and incorporated into the algorithm. This can help Multi-CI to solve real world problems which are generally constrained in nature.

References

1. Civicioglu P (2013) Backtracking search optimization algorithm for numerical optimization problems. Appl Math Comput 219:8121–8144
2. Geem ZW, Kim JH, Loganathan GV (2001) A new heuristicoptimization algorithm: harmony search. Simulation 76(2):60–68
3. Igel C, Hansen N, Roth S (2007) Covariance matrix adaptationfor multi-objective optimization. Evol Comput 15(1):1–28
4. Karaboga D, Akay B (2009) A comparative study of artificial bee colony algorithm. Appl Math Comput 214(1):108–132
5. Karaboga D, Basturk B (2007) A powerful and efficient algorithm for numerical function optimization: artificial bee colony (ABC) algorithm. J Global Optim 39:459–471
6. Kulkarni AJ, Baki MF, Chaouch BA (2016) Application of the cohort-intelligence optimization method to three selected combinatorial optimization problems. Eur J Oper Res 250(2):427–447
7. Kulkarni AJ, Durugkar IP, Kumar M (2013) Cohort intelligence: a self-supervised learning behavior. In IEEE international conference on systems, man, and cybernetics (SMC). pp 1396–1400
8. Li X, Yao X (2012) Cooperatively coevolving particle swarms for large scale optimization. IEEE Trans Evolut Comput 16(2):210–224
9. Liu L, Zhong WM, Qian F (2010) An improved chaos-particle swarm optimization algorithm. J East China Univ Sci Technol (Natural Science Edition) 36:267–272
10. Murugan R, Mohan MR (2012) Modified artificial bee colony algorithm for solving economic dispatch problem. ARPN J Eng ApplSci 7(10):1353–1366
11. Omran MGH, Clerc M (2011) http://www.particleswarm.info/. Accessed 28 June 28 2016
12. Qin AK, Suganthan PN (2005) Self-adaptive differential evolution algorithm for numerical optimization. IEEE Trans Evolut Comput 1(3):1785–1791
13. Selvi V, Umarani R (2010) Comparative analysis of ant colony and particle swarm optimization techniques. Int J Comput Appl 5(4):975–8887
14. Shastri AS, Kulkarni AJ (2018) Multi-cohort intelligence algorithm: an intra-and inter-group learning behaviour based socio-inspired optimization methodology. Int J Parallel Emergent Distrib Syst 33(6):675–715
15. Shastri AS, Nargundkar A, Kulkarni AJ (2020) Multi-cohort intelligence algorithm for solving advanced manufacturing process problems. Neural Comput Appl. https://doi.org/10.1007/s00521-020-04858-y
16. Storn R, Price K (1997) Differential evolution—a simple and efficient heuristic for global optimization over continuous spaces. J Global Optim 11:341–359

17. Suganthan PN, Hansen N, Liang JJ, Deb K, Chen YP, Auger A, Tiwari S (2005) Problem definitions and evaluation criteria forthe CEC 2005 special session on real-parameter optimization. Technical Report 1–50
18. Teo TH, Kulkarni AJ, Kanesan J, Chuah JH, Abraham A (2016) Ideology algorithm: a socio-inspired optimization methodology. Neural Comput Appl. https://doi.org/10.1007/s00521-016-2379-4

Chapter 4
Optimization of Electric Discharge Machining (EDM)

Electric Discharge Machining (EDM) is an electro-thermal, Non-Traditional Machining (NTM) process in which electrical energy is used to generate spark between tool & workpiece and thus material is removed. EDM is mainly used to machine high strength temperature resistant materials and alloys with intricate geometries and is a quite popular NTM process in the machining industry. The process parameters/variables which governs the process are discharge current, gap voltage, pulse on-time, pulse off-time, gap between the work piece and the tool, and dielectric medium. In the process of electrons transfer and erosion during EDM machining, localised heating takes place. This results in case hardening of work piece. Thus, surface roughness becomes important to minimize. This chapter is based on the optimization of process parameters of EDM for minimizing surface roughness and Relative Electrode Wear Rate and maximizing Material Removal Rate using variations of Cohort Intelligence (CI) and Multi-Cohort Intelligence (Multi-CI) [14]. Results achieved using variations of CI and Multi-CI algorithms have been presented and solutions are compared with several contemporary algorithms such as Firefly Algorithm (FA), Genetic Algorithm (GA), Simulated Annealing (SA) and Particle Swarm Optimization (PSO) and Back Propagation Neural Network (BPNN) algorithm. In addition, the performance is compared with the Response Surface Methodology (RSM) approach.

4.1 Introduction

Manufacturing processes can be broadly divided into two groups. The primary manufacturing is related to providing basic shape and size to the material as per design requirements, for example, casting, forming, powder metallurgy, etc. Secondary

© The Editor(s) (if applicable) and The Author(s), under exclusive license to Springer Nature Singapore Pte Ltd. 2021
A. Shastri et al., *Socio-Inspired Optimization Methods for Advanced Manufacturing Processes*, Springer Series in Advanced Manufacturing,
https://doi.org/10.1007/978-981-15-7797-0_4

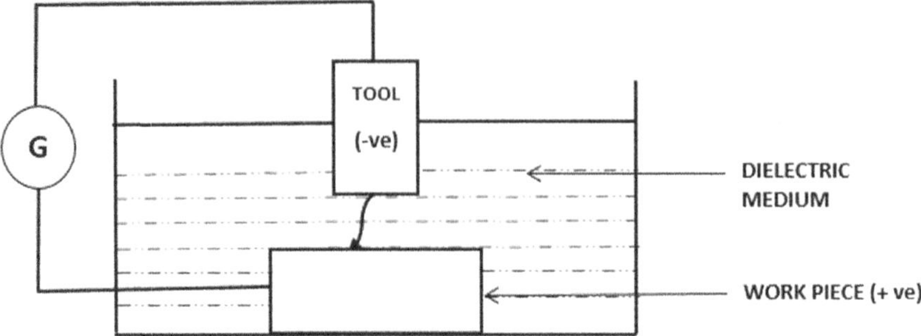

Fig. 4.1 Schematics of EDM

manufacturing processes are also referred to as material removal processes provide final shape and size to the object with necessary dimension control, surface characteristics, etc. They can be further divided into two groups: Conventional Machining Processes and Non-traditional Manufacturing (NTM) Processes. Some notable examples of conventional machining processes are turning, boring, milling, shaping, broaching, slotting, grinding, etc. On the other hand, Ultrasonic Machining (USM), Abrasive Jet Machining (AJM), Electro discharge Machining (EDM) are some of the NTM Processes. New exotic work materials as well as innovative geometric design of products and components are the real motivation behind enriching the capabilities of machining processes to attain desired tolerances with minimum cost. This has led to the development and establishment of NTM processes in the industry as efficient and economic alternatives to conventional ones. The NTM processes have become the prime choice over conventional processes for certain technical requirements [7].

EDM is an electro-thermal NTM in which electrical energy is used to generate spark between tool and workpiece. According to Jain [7], EDM is mainly used to machine high strength temperature resistant materials and alloys with intricate geometries. The material removal in EDM is based on the erosion of unwanted material from the workpiece. Figure 4.1 illustrates the schematics of the EDM process. In EDM, a potential difference is applied between the tool ($-$ve terminal) and work piece (+ve terminal) immersed in a dielectric medium such as kerosene or deionised water. Depending upon the applied potential difference and the gap between the tool and work piece, an electric field is established. The free electrons on the tool are subjected to electrostatic forces and gain a high velocity and energy. These electrons collide with the dielectric molecules which results into ionization. Thus, as the electrons get accelerated, more positive ions and electrons are generated. This cyclic process increases the concentration of electrons and ions in the dielectric medium between the tool and the work piece at the spark gap. Thus, the electrical energy is dissipated as the thermal energy of the spark. According to the experimental investigations by Muthuramalingam and Mohan [11], and Gopalakannan and Senthilvelan [6], during this process electrode wear takes place and hence, electrode wear rate becomes critical to control. The process parameters which essentially govern the process are discharge current, gap voltage, pulse on-time, pulse off-time, gap between the work piece and the tool and dielectric medium. In the process of electrons transfer and erosion, localised heating takes place at the workpiece. This results in case hardening of work piece. Thus, surface roughness R_a becomes important to control.

Das et al. [3] applied Artificial Bee Colony (ABC) algorithm for optimization of *MRR* and surface roughness R_a of EDM. Process modelling and analysis have been attempted by using RSM for EDM job surface integrity by Bhattacharya et al. (2007) to determine the effects of the machining parameters. SA was employed by [19] for optimization of *MRR* and R_a. Analysis of Variance (ANOVA) was applied for determining the contribution of the process parameters for machining metal matrix composites [6]. Continuous Ant Colony Optimization (CACO) was also applied to obtain the optimum parameter setting for *MRR* and R_a [17]. Dewangan et al. [4] investigated the effect of discharge current, pulse on time, tool lift time and tool work time on R_a. Kolli ad Kumar [8] investigated the effects of discharge current, surfactant concentration and powder concentration on the performance of EDM using Taguchi method. Camposeco-Negrete [1] performed experiments on tool steel with EDM for investigating the effect of parameters on R_a and machining time. It was supported by the multi-objective optimization with desirability approach. Experimental investigation and multi response optimization with *MRR* and electrode wear rate as responses of EN8 steel using EDM is carried out by Ganapathy et al. [5]. Dang [2] applied PSO for constrained multi objective optimization of EDM process. Straka and Hašová [16] carried out experimental measurement to identify the impact of selected process parameters on *MRR* and electrode wear in die-sinking EDM of tool steels EN X210Cr12 with Cu electrode EN CW004A. Kumar et al. [10] performed experimentally investigated the effect of various EDM parameters on EDM drilling rate and electrode wear ratio, during the micro drilling of Ti-6Al-7Nb. For comprehensive literature review, refer to the Chap. 1.

A socio inspired optimization technique Cohort intelligence (CI) was developed by Kulkarni et al. in [9]. This algorithm models the activity of the individuals of the society. Basically, it includes certain type of competition and interaction which eventually makes the society to evolve. Further, Patankar and Kulkarni [12] developed variations of CI such as follow better (Fbetter), follow best (Fbest), follow worst, follow itself, follow median, alienation and roulette wheel (RW) (For details, refer Chap. no 2). Multi-CI algorithm (Shastri and Kulkarni [13] is a modified version of CI in which intra and inter cohort learning mechanism amongst different candidates is modelled (For details, refer to the Chap. 3). The problem formulation for EDM process is given in the following section.

4.2 Problem Formulation

As mentioned earlier, the critical process parameters for the EDM process are discharge current (in *A*) b_1, gap voltage (in *V*) b_2, pulse on-time (in *s*) b_3 and pulse off-time (in s) b_4. *MRR*, surface roughness (R_a) and Relative Electrode Wear Rate *REWR* as shown in Eqs. (4.1)–(4.3) are performance responses as they indicate the efficiency of process, surface finish, and electrode wear rate of machined component, respectively. Based on the EDM experiments performed by Tzeng and Chen [18], a regression model is developed by Shukla and Singh [15] and the same is adopted in the current work. The problem formulation is as follows:

$$Maximize\ MRR = -235.15 + 39.7b_1 + 4.227b_2 + 1.569b_3$$
$$- 1.375b_4 - 0.0059b_3^2 - 0.536b_1b_2 \tag{4.1}$$

$$Minimize\ R_a = 31.547 - 0.0618b_1 - 0.438b_2 + 0.059b_3$$
$$- 0.59b_4 + 0.019b_1b_4 - 0.0075b_2b_4 \tag{4.2}$$

$$Minimize\ REWR = 196.564 - 24.19b_1 - 3.135b_2 - 1.781b_3$$
$$+ 0.153b_4 + 0.093b_1^2 + 0.00149b_3^2 + 0.005265b_4^2$$
$$+ 0.464b_1b_2 + 0.158b_1b_3 + 0.025b_1b_4 + 0.029b_2b_3$$
$$- 0.017b_2b_4 - 0.003385b_1b_2b_3 \tag{4.3}$$

where

$$7.5 \le b_1 \le 12.5$$
$$45 \le b_2 \le 55$$
$$50 \le b_3 \le 150$$
$$40 \le b_4 \le 60$$

Sample Code of Roulette Wheel variation is given in the annexure. All other executable codes for all the chapters are given on the following website: sites.google.com/site/oatresearch.

Figure 4.1 illustrates the schematics of the EDM process showing tool as negative terminal and workpiece as positive terminal.

4.3 Numerical Results and Discussion

Multi-CI algorithm [13], variations of CI [12], and PSO algorithm were coded in MATLAB R2017 on Windows Platform with an Intel Core i3 processor and 4 GB RAM. Refer to the appendix for the MATLAB codes of Multi-CI and variations of CI algorithm. For GA and SA algorithms, optimization toolbox is used from the same MATLAB version. The control parameters associated with the Multi-CI, variations

Table 4.1 Control parameters and stopping criteria

Solution methodology	Parameter	Stopping criteria
Multi–CI	No. of cohorts = 3	Objective function value is less than 10^{-16}
	No. of candidates = 5	
	Value of reduction factor = 0.99	
	Behavioral variations for best candidates = 5 and for rest of the candidates = 10	
Variations of CI	No. of candidates = 5	
	Value of reduction factor = 0.99	
SA	Annealing function = Fast annealing	
	Re-annealing interval = 100	
	Temperature update function = Exponential	
	initial temperature = 100	
GA	Population size = 50	
	Scaling function = Rank	
	Selection = Stochastic uniform	
	Crossover fraction = 0.8	
	Mutation = Adaptive feasible	
	Crossover function = Heuristic	
PSO	Inertia co-efficient = Max 0.9 Min 0.2	
	Acceleration co-efficient = 2	

of CI for solving the EDM problem is listed in Table 4.1. Every problem is solved 30 times.

In Table 4.2, best and mean solutions for MRR, R_a and $REWR\%$ of EDM along with associated standard deviation and run time obtained using variations of CI, Multi-CI, GA, SA and PSO are shown. These solutions have been compared with RSM model, BPNN algorithm, and FA as shown in Table 4.3. Since considered EDM problem is inseparable, multimodal and nonlinear in nature, algorithm needs exploration and exploitation abilities to find optimum solution.

Figure 4.2a, b andc shows convergence obtained by different variations of CI such as alienation, follow best, follow better and roulette wheel, Multi-CI, GA and SA, and PSO for objective function MRR, R_a and REW respectively. In follow best and follow better approaches, candidates follow one of the candidates from the cohort. Hence, solutions are getting trapped in local minima as indicated in the convergence plot as learning proceeds, the best and better candidate, jump out of local minima and eventually for all candidates the global minimum is obtained. Unlike follow better and follow best approaches, alienation and roulette wheel approach

Table 4.2 Solutions for MRR, R_a and $REWR$ of EDM

Function		GA	SA	PSO	Roulette wheel	fbest	fbetter	alienation	Multi-CI
MRR	Mean	183.37	182.03	183.37	183.26	38.98	38.24	96.45	183.37
	SD	0.00	2.21	0.00	0.11	0.69	0.71	23.21	0.00
	Best	183.37	183.09	183.37	183.35	39.63	39.52	144.32	183.37
	Run time	1.41	2.61	1.81	0.35	0.59	0.64	0.51	1.81
R_a	Mean	3.55	3.67	3.55	3.61	3.60	3.61	5.99	3.55
	SD	0.00	0.16	0.05	0.03	0.02	0.03	1.76	0.00
	Best	3.55	3.58	3.55	3.55	3.55	3.55	4.06	3.55
	Run time	1.45	2.69	1.89	0.38	0.6	0.63	0.53	1.89
$REWR$	Mean	1.30×10^{-8}	6.82×10^{-4}	3.73×10^{-9}	8.53×10^{-5}	2.96×10^{-5}	7.90×10^{-6}	2.37×10^{-7}	1.85×10^{-8}
	S.D.	0.00	0.05	0.00	0.00	0.00	0.00	0.00	0.00
	Best	1.22×10^{-7}	1.36×10^{-2}	8.95×10^{-9}	2.43×10^{-4}	1.49×10^{-4}	4.15×10^{-5}	9.79×10^{-7}	9.50×10^{-9}
	Run time	1.7	2.8	1.42	0.42	0.62	0.66	0.57	1.93

Mean Mean solution *SD* Standard-deviation *Best* Best solution *Runtime* Mean runtime in seconds

Table 4.3 Comparison of solutions for solving MRR, R_a and $REWR$ of EDM

Problem	RSM Tzeng and Chen [18]	BPNN Tzeng and Chen [18]	FA Shukla and Singh [15] b	Roulette wheel	fbest	fbetter	Alienation	GA	SA	PSO	Multi-CI
MRR	157.39	159.70	181.67	183.35	39.63	39.52	144.32	183.35	183.09	183.37	183.37
R_a	7.38	7.04	3.67	3.55	3.55	3.55	4.06	3.55	3.58	3.55	3.55
REWR %	7.63	6.21	6.3250×10^{-5}	2.4350×10^{-4}	2.9650×10^{-5}	7.9050×10^{-6}	9.7950×10^{-9}	1.2250×10^{-7}	2.4350×10^{-4}	1.8550×10^{-9}	9.50×10^{-9}

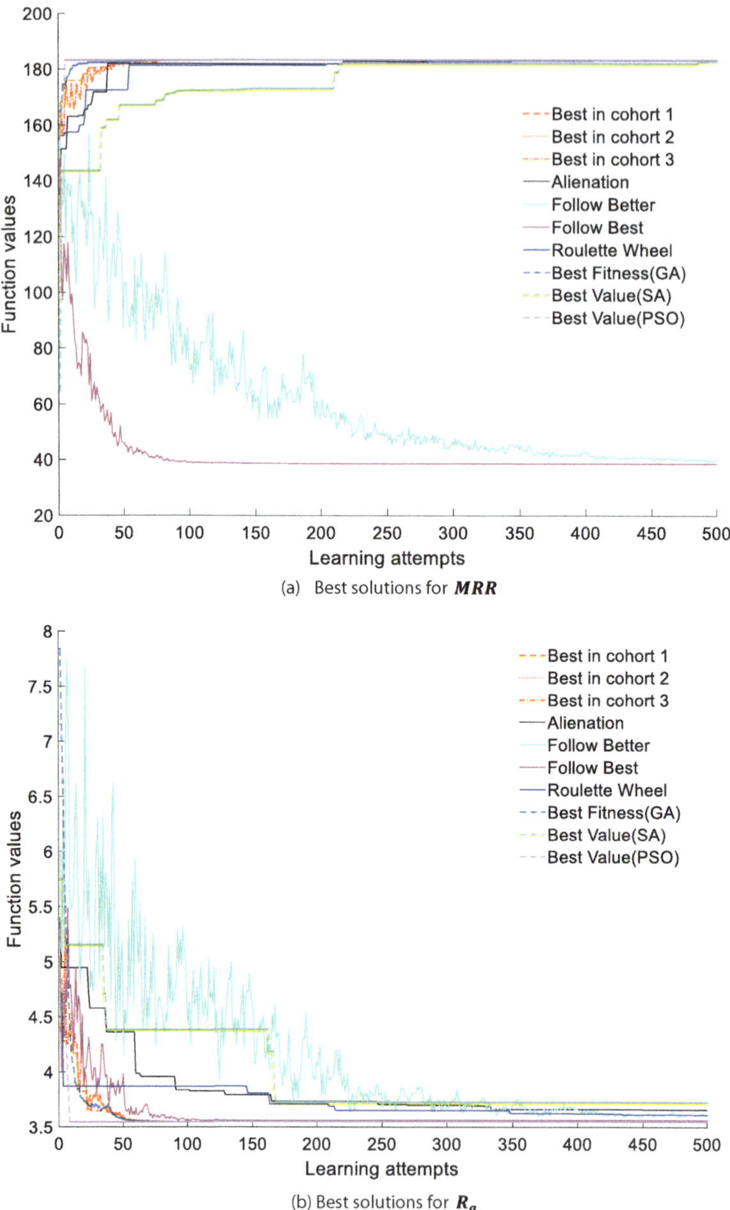

(a) Best solutions for $\textbf{\textit{MRR}}$

(b) Best solutions for $\textbf{\textit{R}}_{\textbf{\textit{a}}}$

Fig. 4.2 Convergence plots for EDM

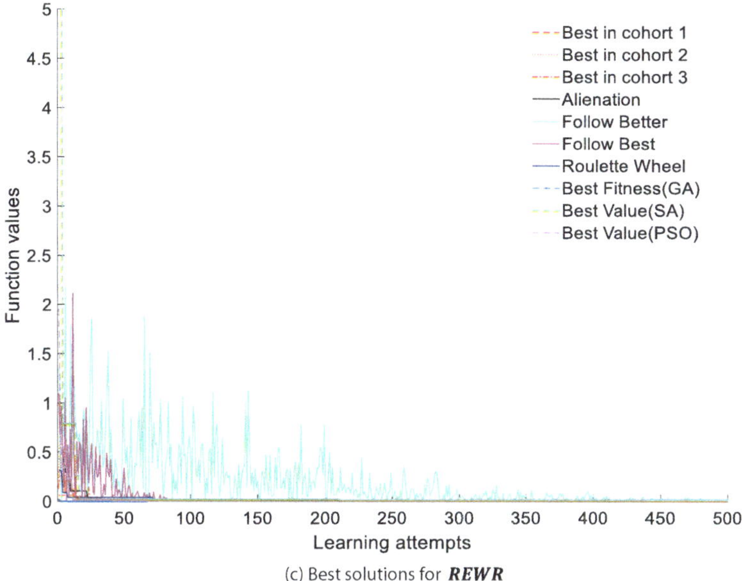

(c) Best solutions for *REWR*

Fig. 4.2   (continued)

have shown significant difference in follow mechanism and results in global solution without trapping into local minima. GA is population based optimization technique. Since a population is updated in every generation, a set of solutions can be obtained simultaneously. This gives exploration power to GA. GA exploits probabilistic rules to further guide the search. When population converges at some location in decision space, the search direction narrows down and a near-optimal solution is found. This generally is resulted in comparatively less robustness and higher standard deviation. It is evident from the solutions presented in Table 4.3. SA imitates the annealing process used in metallurgy. SA begins with the high temperature referred to as initial temperature with which search space is explored. As the temperature declines, the algorithm exploits certain space and settles at a particular point or a local optimum. A temperature update function generates possible states to be explored. To find global optimum, SA requires careful controlling and fine tuning of the above mentioned parameters and hence it might not result in finding global optimum. SA has resulted in the worst solution quality for the objective functions as compared with GA, PSO, Variations of CI and Multi-CI algorithms as shown in Table 4.3. The Multi-CI algorithm performed exceedingly better as compared with the RSM, BPNN solutions [18], and SA [15] as evident from Table 4.3. In Multi-CI exploration and exploitation mechanism is strong as every candidate search for the best solution from its own cohort as well as other cohorts. Due to this there are less chances of getting trapped into local minima which results in the faster convergence. For EDM, 47–80% maximization of Material Removal Rate; 2–13% and 92–98% minimization of surface roughness and Relative Electrode Wear Rate, respectively have been attained.

4.4 Summary

In this work, four variations of CI algorithm, Multi-CI, GA, SA and PSO have been successfully applied on the minimization of surface roughness and electrode wear rate and maximization of material removal rate of EDM process. Solutions were compared with GA, SA, PSO, RSM, FA, and BPNN algorithms. Maximum material removal rate has been predicted by Multi-CI whereas for surface roughness and electrode wear rate, optimum results have been obtained using GA, variations of CI and Multi-CI. GA and the approach of Multi-CI have proved to be robust for solving material removal rate and surface roughness problems for EDM process. In near future, authors intend to apply CI and Multi-CI for more complex and constrained manufacturing process optimization problems. The next chapter is based on the optimization of process parameters for Abrasive Water Jet Machining to minimize surface roughness and $kerf$.

References

1. Camposeco-Negrete C (2019) Prediction and optimization of machining time and surface roughness of AISI O1 tool steel in wire-cut EDM using robust design and desirability approach. Int J Adv Manuf Technol 1–12
2. Dang XP (2018) Constrained multi-objective optimization of EDM process parameters using kriging model and particle swarm algorithm. Mater Manuf Processes 33(4):397–404
3. Das MK, Kumar K, Barman TK, Sahoo P (2014) Application of artificial bee colony algorithm for optimization of MRR and surface roughness in EDM of EN31 tool steel. Procedia Mater Sci 6:741–751
4. Dewangan S, Gangopadhyay S, Biswas CK (2015) Multi-response optimization of surface integrity characteristics of EDM process using grey-fuzzy logic-based hybrid approach. Eng Sci Technol Int J 18(3):361–368
5. Ganapathy S, Balasubramanian P, Senthilvelan T, Kumar R (2019) Multi-response optimization of machining parameters in EDM using square-shaped nonferrous electrode. In: Advances in manufacturing processes. Springer, Singapore, pp 287–295
6. Gopalakannan S, Senthilvelan T (2014) Optimization of machining parameters for EDM operations based on central composite design and desirability approach. J Mech Sci Technol 28(3):1045–1053
7. Jain VK (2008) Advanced (non-traditional) machining processes. In: Machining. Springer, London. pp 299–327
8. Kolli M, Kumar A (2015) Effect of dielectric fluid with surfactant and graphite powder on electrical discharge machining of titanium alloy using Taguchi method. Eng Sci Technol Int J 18(4):524–535
9. Kulkarni AJ Durugkar IP Kumar M (2013) Cohort intelligence: a self supervised learning behavior. In: 2013 IEEE international conference on systems, man, and cybernetics. IEEE, pp 1396–1400
10. Kumar, K., Singh, V., Katyal, P. and Sharma, N., 2019. EDM μ-drilling in Ti-6Al-7Nb: experimental investigation and optimization using NSGA-II. Int J Adv Manuf Technol pp 1–12
11. Muthuramalingam T, Mohan B (2015) A review on influence of electrical process parameters in EDM process. Arch Civil Mech Eng 15(1):87–94

12. Patankar NS, Kulkarni AJ (2018) Variations of cohort intelligence. Soft Comput 22(6):1731–1747

13. Shastri AS, Kulkarni AJ (2018) Multi-cohort intelligence algorithm: an intra-and inter-group learning behaviour based socio-inspired optimisation methodology. Int J Parallel Emergent Distrib Syst 33(6):675–715

14. Shastri AS, Nargundkar A, Kulkarni AJ (2020) Multi-Cohort intelligence algorithm for solving advanced manufacturing process problems. Neural Comput Appl. https://doi.org/10.1007/s00521-020-04858-y

15. Shukla R, Singh D (2017) Selection of parameters for advanced machining processes using firefly algorithm. Eng Sci Technol Int J 20(1):212–221

16. Straka LU, Hašová S (2018) Prediction of the heat-affected zone of tool steel EN X37CrMoV5-1 after die-sinking electrical discharge machining. Proc Inst Mech Eng Part B J Eng Manuf 232(8):1395–1406

17. Teimouri R, Baseri H (2014) Optimization of magnetic field assisted EDM using the continuous ACO algorithm. Appl Soft Comput 14:381–389

18. Tzeng CJ, Chen RY (2013) Optimization of electric discharge machining process using the response surface methodology and genetic algorithm approach. Int J Prec Eng Manuf 14(5):709–717

19. Yang SH, Srinivas J, Mohan S, Lee DM, Balaji S (2009) Optimization of electric discharge machining using simulated annealing. J Mater Process Technol 209(9):4471–4475

Chapter 5
Optimization of Abrasive Water Jet Machining (AWJM)

Abrasive Water Jet Machining (AWJM) is an advanced version of Abrasive Jet Machining (AWJ) which employs water as the carrier medium for abrasive particles. The AWJM process can machine complex shapes and importantly, doesn't generate heat concentrated zones. Work piece thickness, nozzle diameter, standoff distance and traverse speed are the typical process parameters/variables for AWJM. Kerf taper angle and surface roughness are performance responses as they indicate the geometry and surface finish of machined component, respectively. This chapter is based on the optimization of process parameters for AWJM to minimize surface roughness and *kerf* using socio inspired optimization algorithm Multi-Cohort Intelligence (Multi-CI) [18]. Results achieved have been presented and solutions are compared with experimental results, regression model, variations of Cohort Intelligence (CI), Firefly Algorithm (FA), Genetic Algorithm (GA), Simulated Annealing (SA), and Particle Swarm Optimization (PSO). Multi-CI algorithm has resulted in significantly better solutions.

5.1 Introduction

The advancement of materials science during the last few decades has led to the development of many hard-to-machine materials, such as titanium, stainless steel, high-strength temperature-resistant alloys, ceramics, refractories, fibre-reinforced composites, super alloys, etc. These materials are not suitable to be machined by the conventional machining processes because of their high hardness, strength, brittleness, toughness and low machinability properties. Hence, newer processes in machining, known as non-traditional machining (NTM) processes have been developed [8]. These are non-traditional in the sense that the conventional cutting tools are

A. Shastri et al., *Socio-Inspired Optimization Methods for Advanced
Manufacturing Processes*, Springer Series in Advanced Manufacturing,
https://doi.org/10.1007/978-981-15-7797-0_5

not employed in these machining processes for metal removal; instead energy in its direct form is utilized. Some notable examples of conventional machining processes are turning, boring, milling, shaping, broaching, slotting, grinding, etc. On the other hand, Ultrasonic Machining (USM), Abrasive Jet Machining (AJM), Abrasive Water Jet Machining (AWJM), Electro discharge Machining (EDM) are some of the NTM Processes [14].

In a generalized AJM process very high velocity fine abrasive particles impinge on the work material resulting into cutting of the workpiece. This process is very effective method for hard and brittle materials [7]. AWJM is an advanced version of AJM which employs water as the carrier medium for abrasive particles. The AWJM process can machine complex shapes and importantly, does not generate heat concentrated zones [12]. The AWJM can be used for machining of a wide variety of materials including advanced layered composites. Further development of abrasives leads to AWJM capability improvement [1]. According to Kechagias et al. [9], Shanmugam et al. [16] and Shukla and Singh [20] there are four significant process parameters, viz. workpiece thickness, nozzle diameter, distance between cutting head and workpiece referred to as standoff distance, and speed of cutting head over the workpiece referred to as traverse speed (refer to Fig. 5.1). The associated responses are surface roughness R_a and taper angle. Refer to Fig. 5.2 for the illustration of $kerf$ geometry [2]. It shows a typical cut produced by AWJM. Ideally, the top width and bottom width of the cut should necessarily be equal; however, AWJM produces a tapered cut and it is necessary to minimize this taper angle. Along with the $kerf$ geometry, another significant process criterion is internal surface roughness R_a. It plays an important role while assembling the parts as it may directly affect the required fit and functionality. Furthermore, as the nozzle diameter increases with respect to the thickness of the workpiece, the $kerf$ top width increases, however, the depth of cut decreases as abrasive particles cannot reach the bottom of the workpiece. Thus, the thickness of the workpiece also plays a critical role along with nozzle diameter for getting optimized $kerf$ value. The distance between the nozzle exit and the surface of the workpiece is referred to as the standoff distance. Any increase in the standoff distance reduces abrasive particles velocity which further decreases the impinged water pressure on the workpiece. This results in gradual reduction of depth of cut along the workpiece thickness and in turn a tapered cut is produced. Hence, it is important to optimize the standoff distance. The velocity at which the nozzle moves over the workpiece surface is referred to as traverse speed. If traverse speed is too high, the stream of water abrasive particles does not get sufficient time to cut the material and this results in poor erosion. This highlights that the AWJM process parameters interact with one another and affects the process responses $kerf$ and R_a. This necessitate identifying an optimum combination of the process parameters. For comprehensive literature review, refer to the Chap. 1.

Several nature inspired algorithms have been applied to obtain optimized combination of the AWJM process parameters. [19] carried out AWJM experimentation using Taguchi method and compared with FA as well as evolutionary algorithms such as PSO [3], Black Hole Algorithm (BH) [6], Cuckoo Search Algorithm [22], SA [10] and Bio-Geography based Optimization (BBO) [21]. Furthermore, [20]

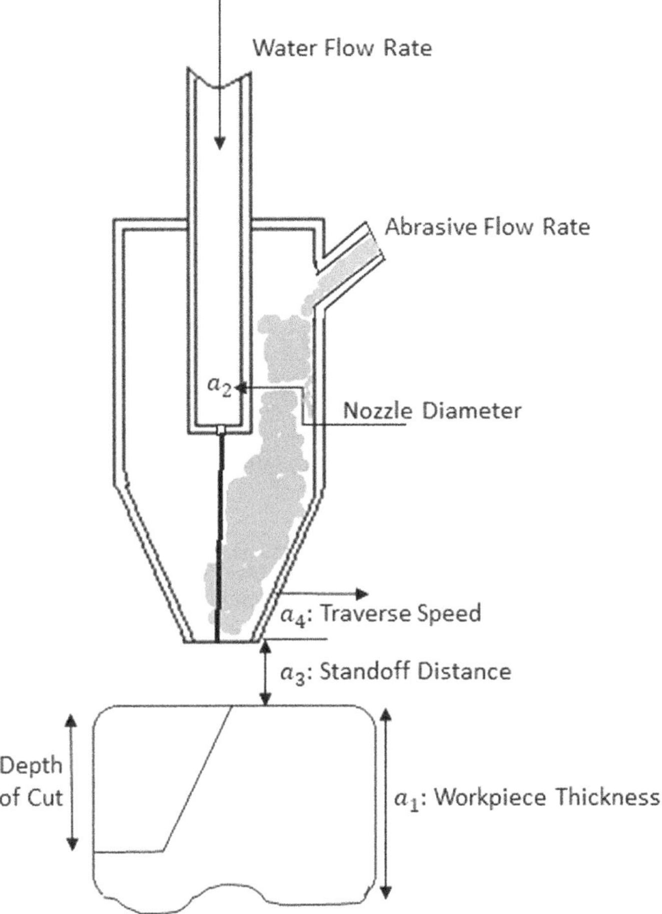

Fig. 5.1 AWJM process and parameters

applied FA to EDM and AWJM. Schwartzentruber et al. [15] used GA to optimize a nozzle design and operating conditions for AWJM in order to maximize the abrasive energies. Jain et al. [8] applied GA for optimization of process parameters of four advanced machining processes including AWJM. Gostimirovic et al. [4] investigated the trajectory profile generation in AWJM which was modelled using Genetic Programming (GP) for the jet lag, angle deviation and radius of the trajectory curvature in the cutting zone. Zain et al. [24] applied an integrated SA-GA algorithm to obtain optimal solution of AWJM process parameters yielding minimum value of surface roughness R_a.

A socio inspired optimization technique Cohort intelligence (CI) was developed by Kulkarni et al. in [11]. This algorithm models the activity of the individuals of the society. Basically, it includes certain type of competition and interaction which

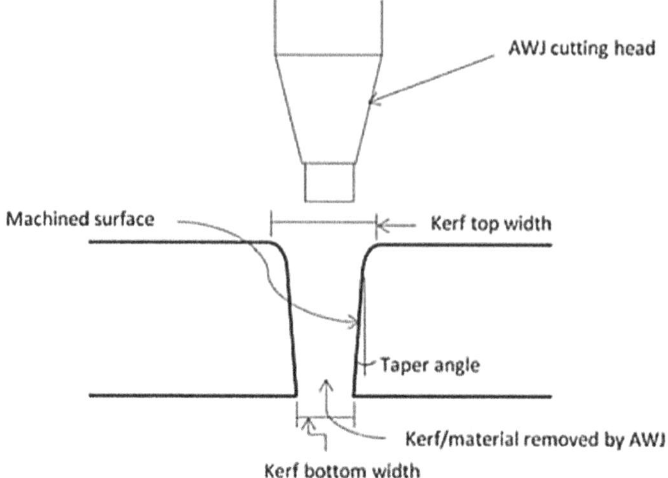

Fig. 5.2 AWJM *kerf* Illustration [2]

eventually makes the society to evolve. Further, Patankar and Kulkarni [13] developed variations of CI such as follow better (Fbetter), follow best (Fbest), follow worst, follow itself, follow median, alienation and roulette wheel (RW) (For details, refer Chap. 2). Multi-CI algorithm [17] is a modified version of CI in which intra and inter cohort learning mechanism amongst different candidates is modelled (For details, refer to the Chap. 3). The problem formulation for AWJM process is given in the following section.

5.2 Problem Formulation

As mentioned earlier, the critical process parameters for the AWJM process are work piece thickness (in *mm*) a_1, nozzle diameter (in *mm*) a_2, standoff distance (in *mm*) a_3 and traverse speed (in *mm/min*) a_4. *kerf* and surface roughness (R_a) are performance responses as they indicate the geometry and surface finish of machined component, respectively. Based on the AWJM experiments performed by Kechagias et al. [9], a regression model is developed by Shukla and Singh [20] is adopted in the current work and is shown in Eqs. (5.1) and (5.2) respectively. The standoff distance a_3 and traverse speed a_4 have relatively larger search spaces as compared to the rest two variables. The problem formulation is as follows:

$$\begin{aligned}
Minimize\ R_a = &-23.309555 + 16.6968a_1 + 26.9296a_2 + 0.0587a_3 \\
&+ 0.0146a_4 - 5.1863a_2^2 - 10.4571a_1a_2 - 0.0534a_1a_3 - 0.0103a_1a_4 \\
&+ 0.0113a_2a_3 - 0.0039a_2a_4
\end{aligned} \tag{5.1}$$

$$\begin{aligned}
Minimize\ kerf\ = &-1.15146 + 0.70118a_1 + 2.72749a_2 + 0.00689a_3 \\
&- 0.00025a_4 + 0.00386a_2a_3 - 0.93947a_2^2 - 0.25711a_1a_2 - 0.00314a_1a_3 \\
&- 0.00249a_1a_4 + 0.00196a_2a_4 - 0.00002a_3a_4 - 0.00001a_3^2
\end{aligned} \tag{5.2}$$

where

$$0.9 \leq a_1 \leq 1.25$$
$$0.95 \leq a_2 \leq 1.5$$
$$20 \leq a_3 \leq 96$$
$$200 \leq a_4 \leq 600$$

Sample Code of Roulette Wheel variation for AWJM is given in the annexure. All other executable codes for all the chapters are given on the following website: https://sites.google.com/site/oatresearch.

5.3   Numerical Results and Discussion

Multi-CI, and PSO algorithm were coded in MATLAB R2017 on Windows Platform with an Intel Core i3 processor and 4 GB RAM. Refer to the appendix section for the MATLAB codes of Multi-CI and variations of CI algorithm. For GA and SA algorithms, optimization toolbox is used from the same MATLAB version. The control parameters associated with the Multi-CI, SA, GA, and PSO for solving the AWJM problem is listed in Table 5.1. To make the results statistically significant and robust, every problem is solved 30 times.

In Table 5.2, best and mean solutions for $\mathbf{R_a}$ and **kerf** of AWJM along with associated standard deviation and run time obtained using Multi-CI, GA, SA and PSO are presented. These solutions have been compared with variations of CI, SA, FA, experimental and regression results as shown in Table 5.3. Since considered AWJM problem is inseparable, nonlinear with quadratic terms, algorithm needs better exploration and exploitation mechanism to find optimum solution without getting trapped into local minima.

In FA [23] light intensity is directly proportional to the distance between two fireflies. Best Firefly position gets updated based on current position, desirability to another beautiful firefly, and product of randomization constraint and the random number. Therefore, in nonlinear optimization problem algorithm shows limitation of getting stuck into local minimum and sometimes shows premature convergence as compared to other algorithms. According to Shukla and Singh [20] the results obtained using FA are better when compared with experimental results. In Multi-CI exploration and exploitation mechanism is strong as every candidate search for the best solution from its own cohort as well as other cohorts. Due to this there are less chances of getting trapped into local minima which results in the faster

Table 5.1 Control parameters and stopping criteria

Solution methodology	Parameter	Stopping criteria
Multi-CI	No. of cohorts = 3	Objective function value is less than 10^{-16}
	No. of candidates = 5	
	Value of reduction factor = 0.99	
	Behavioral variations for best candidates = 5 and for rest of the candidates = 10	
SA	Annealing function = Fast annealing	
	Re-annealing interval = 100	
	Temperature update function = Exponential	
	Initial temperature = 100	
GA	Population size = 50	
	Scaling function = Rank	
	Selection = Stochastic uniform	
	Crossover fraction = 0.8	
	Mutation = Adaptive feasible	
	Crossover function = Heuristic	
PSO	Inertia co-efficient = Max 0.9 Min 0.2	
	Acceleration co-efficient = 2	

Table 5.2 Solutions for R_a and $kerf$ of AWJM using Multi-CI, GA, SA

Function		GA	SA	PSO	Multi-CI
R_a	Mean	4.43	4.86	4.39	4.38
	SD	0.03	0.12	0.22	0.00
	Best	4.38	4.61	4.75	4.38
	Run time	1.62	2.63	1.78	4.63
$kerf$	Mean	0.33	0.41	0.43	0.33
	SD	0.01	0.04	0.00	0.01
	Best	0.33	0.36	0.43	0.33
	Run time	1.48	2.8	1.52	3.89

Mean Mean solution; *SD* Standard-deviation; *Best* Best solution; *Runtime* Mean runtime in seconds

convergence. Multi-CI outperformed FA in terms of solution quality solving AWJM problem as evident in Tables 5.2 and 5.3. SA has resulted in the worst solution quality for the objective functions as compared with GA, Variations of CI and Multi-CI algorithms as shown in Table 5.3. Whereas, solutions with PSO are comparable

Table 5.3 Comparison of Algorithms for solving R_a and $kerf$ of AWJM

Problem	Experimental Kechagias [9]	Regression Kechagias [9]	FA Shukla and Singh [19]	Roulette wheel Gulia and Nargundkar [5]	fbest	fbetter	Alienation	GA	SA	PSO	Multi-CI
R_a	5.80	5.41	4.44	4.38	4.38	4.38	4.38	4.38	4.61	4.39	4.38
kerf	0.85	0.90	0.37	0.34	0.34	0.34	0.34	0.33	0.35	0.43	0.33

in case of surface roughness but worse when evaluated for $kerf$. In addition, Multi-CI performed exceedingly better when compared with experimental and regression solutions [9]. Figure 5.3a and b shows convergence obtained by Multi-CI, GA, SA

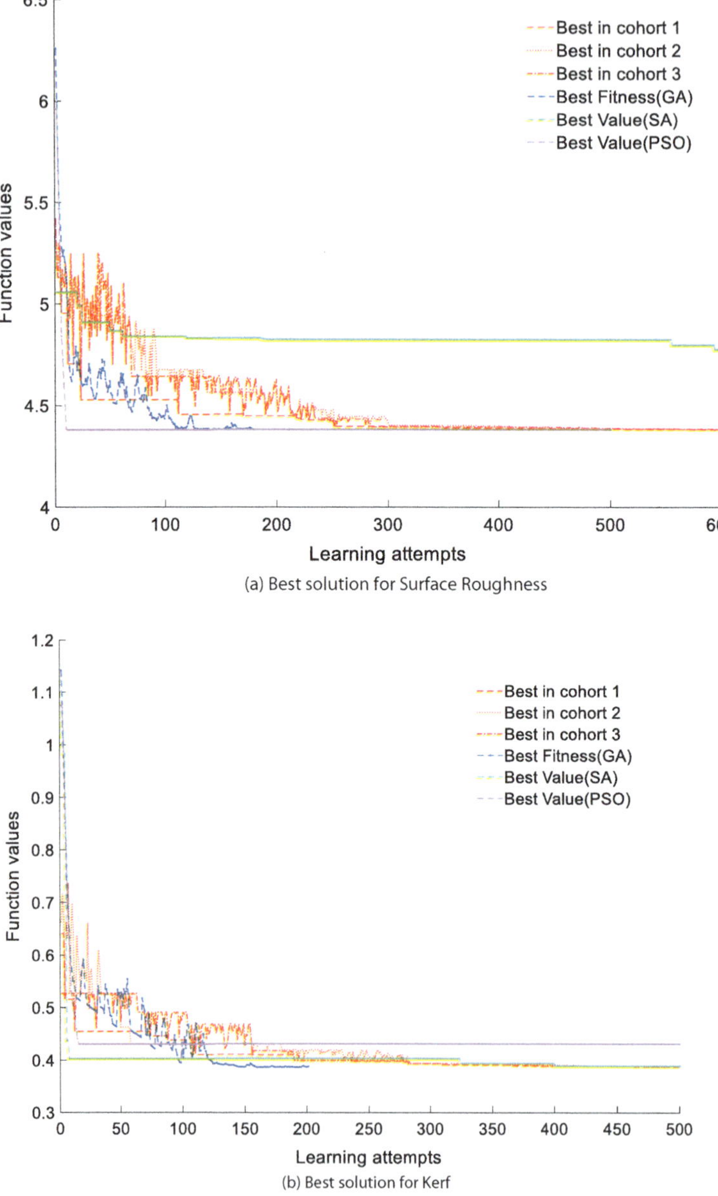

(a) Best solution for Surface Roughness

(b) Best solution for Kerf

Fig. 5.3 Convergence plot for AWJM process

and PSO for objective functions surface roughness and $kerf$ respectively. It is evident in Tables 5.2 and 5.3, for AWJM Multi-CI achieved 5% and 8% minimization of R_a, as well as 8% and 23% minimization of $kerf$ when compared with SA and PSO.

5.4 Summary

In this chapter, Multi-CI, GA, SA, and PSO have been successfully applied for minimizing surface roughness and $kerf$ of AWJM process. Solutions are compared with variations of CI, FA, experimental results and regression results. Multi-CI outperformed experimental, regression, and FA solutions whereas, results are comparable with the variations of CI and GA. Overall, Multi-CI exhibited excellent performance against FA, SA, PSO, experimental and regression results solving the complex AWJM process problems. In near future, authors intend to apply CI and Multi-CI algorithms to more complex and constrained AWJM process optimization problems. In the next chapter performance parameters associated with micro milling process viz. surface roughness and machining time have been optimized.

References

1. Armağan M, Arici AA (2017) Cutting performance of glass-vinyl ester composite by abrasive water jet. Mater Manuf Process 32(15):1715–1722
2. Dhanawade A, Kumar S, Kalmekar RV (2016) Abrasive water jet machining of carbon epoxy composite. Def Sci J 66(5):522–528
3. Eberhart R, Kennedy J (1995) A new optimizer using particle swarm theory. In: MHS'95, Proceedings of the sixth international symposium on micro machine and human science. IEEE, pp 39–43
4. Gostimirovic M, Pucovsky V, Sekulic M, Rodic D, Pejic V (2019) Evolutionary optimization of jet lag in the abrasive water jet machining. Int J Adv Manuf Technol 101(9–12):3131–3141
5. Gulia V, Nargundkar A (2019) Optimization of process parameters of abrasive water jet machining using variations of cohort intelligence (CI). In: Applications of artificial intelligence techniques in engineering. Springer, Singapore, pp 467–474
6. Hatamlou A (2013) Black hole: a new heuristic optimization approach for data clustering. Inf Sci 222:175–184
7. Jagadeesh T (2015) Non traditional machining. Mechanical Engineering Department, National Institute of Technology, Calicut
8. Jain VK (2008) Advanced (non-traditional) machining processes. In: Machining. Springer, London, pp 299–327
9. Kechagias J, Petropoulos G, Vaxevanidis N (2012) Application of Taguchi design for quality characterization of abrasive water jet machining of TRIP sheet steels. Int J AdvManuf Technol 62(5–8):635–643
10. Kirkpatrick S, Gelatt CD, Vecchi MP (1983) Optimization by simulated annealing. Science 220(4598):671–680
11. Kulkarni AJ, Durugkar IP, Kumar M (2013) Cohort intelligence: a self supervised learning behavior. In: 2013 IEEE international conference on systems, man, and cybernetics. IEEE, pp 1396–1400

12. Momber AW, Kovacevic R (2012) Principles of abrasive water jet machining. Springer Science & Business Media
13. Patankar NS, Kulkarni AJ (2018) Variations of cohort intelligence. Soft Comput 22(6):1731–1747
14. Samanta S, Chakraborty S (2011) Parametric optimization of some non-traditional machining processes using artificial bee colony algorithm. Eng Appl Artif Intell 24(6):946–957
15. Schwartzentruber J, Narayanan C, Papini M, Liu HT (2016) Optimized abrasive waterjet nozzle design using genetic algorithms. In: The 23rd international conference on water jetting. At Seattle, USA
16. Shanmugam DK, Nguyen T, Wang J (2008) A study of delamination on graphite/epoxy composites in abrasive waterjet machining. Compos A Appl Sci Manuf 39(6):923–929
17. Shastri AS, Kulkarni AJ (2018) Multi-cohort intelligence algorithm: an intra-and inter-group learning behaviour based socio-inspired optimisation methodology. Int J Parallel Emergent Distrib Syst 33(6):675–715
18. Shastri AS, Nargundkar A, Kulkarni AJ (2020) Multi-Cohort intelligence algorithm for solving advanced manufacturing process problems. Neural Comput Appl. https://doi.org/10.1007/s00521-020-04858-y
19. Shukla R, Singh D (2017) Experimentation investigation of abrasive water jet machining parameters using Taguchi and evolutionary optimization techniques. Swarm and Evolut Comput 32:167–183
20. Shukla R, Singh D (2017) Selection of parameters for advanced machining processes using firefly algorithm. Eng Sci Technol Int Jl 20(1):212–221
21. Simon D (2008) Biogeography-based optimization. IEEE Trans Evol Comput 12(6):702–713
22. Yang XS, Deb S (2009) Cuckoo search via Lévy flights. In: 2009 world congress on nature & biologically inspired computing (NaBIC). IEEE, pp 210–214
23. Yang XS, (2009) Firefly algorithms for multimodal optimization. In: International symposium on stochastic algorithms. Springer, Berlin, Heidelberg. pp 169–178
24. Zain AM, Haron H, Sharif S (2011) Estimation of the minimum machining performance in the abrasive water jet machining using integrated ANN-SA. Expert Syst Appl 38(7):8316–8326

Chapter 6
Optimization of Micro Milling Process

Micro-Milling refers to a basic end-milling process using tools up to 1 mm in diameter. The geometry that can be produced by micro-end-milling is more flexible than those produced by lithography and other traditional micro manufacturing techniques. Furthermore, a wide range of materials could be processed using micro end milling. This chapter is based on the optimization of process parameters of micro milling performed on polymethyl methacrylate (PMMA) workpiece. Two pertinent milling cutter diameters, viz. 0.7 and 1 mm have been used for minimization of surface roughness and machining time. Socio inspired optimization methodologies Variations of Cohort Intelligence (CI) and Multi-Cohort Intelligence (Multi-CI) [13] have been applied to optimize the regression model developed with the experimentation work. Results achieved have been presented and solutions are compared with Genetic Algorithm (GA), Simulated Annealing (SA), and Particle Swarm Optimization (PSO).

6.1 Introduction

Manufacturing processes can be broadly divided into two groups. The primary manufacturing is related to providing basic shape and size to the material as per design requirements, for example, casting processes, forming processes, powder metallurgical processes, etc. Secondary manufacturing processes are referred to as material removal processes. They provide final shape and size to the object with necessary dimension control, surface characteristics, etc. They can be further divided into two groups: Conventional Machining processes and Non-traditional Manufacturing (NTM) processes [5]. Some notable examples of conventional machining processes are turning, boring, milling, shaping, broaching, slotting, grinding, etc.

A. Shastri et al., *Socio-Inspired Optimization Methods for Advanced Manufacturing Processes*, Springer Series in Advanced Manufacturing, https://doi.org/10.1007/978-981-15-7797-0_6

On the other hand, Ultrasonic Machining (USM), Abrasive Jet Machining (AJM), Electro discharge Machining (EDM) are some of the NTM processes [5]. New exotic work materials as well as innovative geometric design of components are the real motivation behind enriching the capabilities of machining processes to attain desired tolerances with minimum cost. This has further led to the development and establishment of the NTM processes as efficient and economic alternatives to conventional ones. The NTM processes have become the prime choice over conventional processes for certain technical requirements [5].

The conventional machining processes have sufficed the requirements of the industries over the decades; however, the need for meso (500 μm–10 mm) and micro scale (1–500 μm) products are hastily increasing in various fields such as aeronautical, biomedical, automobile, optical and nuclear and semiconductor sector. The associated materials are generally ceramics, metals, polymers, etc. Micro manufacturing is the key technology that can ensure the realization of miniature features in a product. Mechanical micromachining is further classified into conventional tool based and advanced micromachining/nano finishing techniques. Conventional tool based micromachining techniques include micro turning, micro drilling and micro milling processes. There are several parameters associated with these processes which need to be controlled in order to achieve the desired responses.

Milling is a machining process in which metal is removed by a rotating multiple-tooth cutter. Each tooth removes a small amount of metal with each revolution of the spindle. An end mill is a milling cutter, which is shank-mounted to the machine tool. It has cutting edges on the face end as well as on the periphery and may be single or double End construction. End mills are the most common and widely used type of milling cutters. Micro-end-milling refers to a basic end milling process using tools less than 1 mm in diameter shown in Fig. 6.1. The geometry that can be produced by

Fig. 6.1 Milling process and parameters [4]

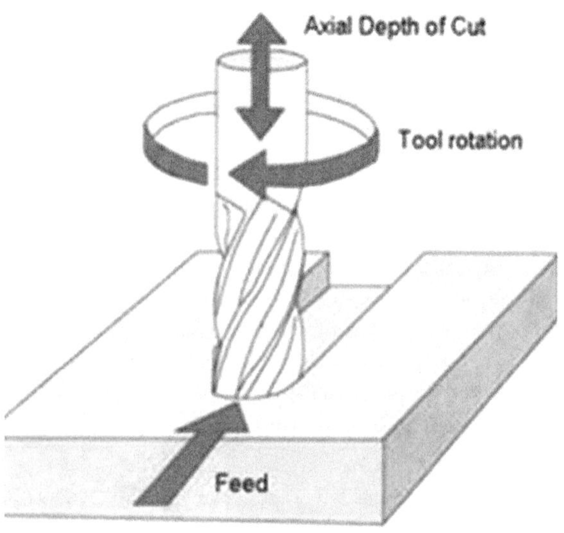

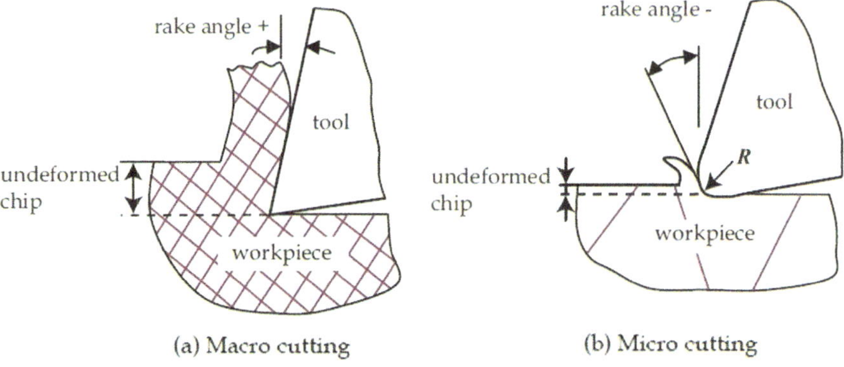

Fig. 6.2 Cutting mechanism in micro-milling [8]

micro-end-milling are more flexible than those produced by lithography and other traditional techniques. Furthermore, wide range of materials can be processed using this process. It is important for the production of meso-scale parts (parts in the order of 1 mm to 1 cm) which are too large for lithography techniques, but too small for many other traditional processing techniques. The basic characteristic of the micro-milling process is essentially similar to that of the conventional milling process with the main difference being in the size of the tool used and cutting conditions.

Micro cutting is significantly influenced by size effect. With the scaling down of cutting geometry, the applicable machining parameters in micro cutting are comparable in magnitude to the cutting-edge radius of micro cutting tool and grain size of work piece materials, which in turn affects chip formation, tool performance, surface formation, etc. The size effect is illustrated in Fig. 6.2 which shows the interaction between the cutting tool and work piece in orthogonal cutting, both in micro scale and macro scale. For comprehensive literature review, refer to the Chap. 1.

Several nature inspired algorithms have been applied on micro milling process to obtain optimized combination of the parameters. Rehman et al. [10] studied the failure mechanism and factors which affect the micro end milling during machining of pure copper. Gopalasamy et al. [3] applied Taguchi method to find optimum process parameters for end milling. Zain et al. [14] optimized the effect of the radial rake angle of the tool combined with cutting speed and feed rate for the surface roughness with GA. Sarvanan et al. [11] performed multi-objective optimization of MRR and tool eccentricity with GA considering feed rate and torque as process variables. Kumar et al. [7] experimented micro milling of PMMA biomaterial and optimized the parameters with GA for CAPP applications. Pham et al. [2] investigated the parameters affecting the roughness of surfaces produced in the micro milling process for two different materials Aluminium and Copper. ANOVA was performed further. Bao et al. [1] had proposed Genetically Optimized Neural Network System (GONNS) for the selection of optimal cutting conditions from the experimental data when analytical or empirical mathematical models are not available.

A socio inspired optimization technique Cohort intelligence (CI) was developed by Kulkarni et al. in [6]. This algorithm models the activity of the individuals of the society. Basically, it includes certain type of competition and interaction which eventually makes the society to evolve. Further, Patankar and Kulkarni [9] developed variations of CI such as follow better (Fbetter), follow best (Fbest), follow worst, follow itself, follow median, alienation and roulette wheel (RW) (For details, refer Chap. 2). Multi-CI algorithm [12] is a modified version of CI in which intra and inter cohort learning mechanism amongst different candidates is modelled (For details, refer to the Chap. 3). The problem formulation for Micro Milling process is given in the following section.

6.2 Problem Formulation

As mentioned earlier, the critical process parameters for the micro milling process are cutting speed (d_1 in rpm), and feed (d_2 in $\frac{mm}{min}$). Surface roughness (R_a) and machining time (m_t) are performance responses as they indicate the surface finish of machined component and productivity, respectively. The micro milling experiments performed by Kumar et al. [7], and a regression model developed, is adopted in the current work and is shown in Eqs. (6.1), (6.2), (6.3) and (6.4) respectively. The problem formulation is as follows:

For tool diameter 0.7 mm

$$Minimize\ R_a = -0.455378 + 0.00027d_1 + 0.016422d_2 - 0.000077d_1d_2 \quad (6.1)$$

$$Minimize\ M_t = 17.7164 - 0.0002d_1 - 4.8404d_2 + 0.0001d_1d_2 \quad (6.2)$$

For tool diameter 1 mm

$$Minimize\ R_a = -0.208871 + 0.000144d_1 + 0.019571d_2 \quad (6.3)$$

$$Minimize\ M_t = 20.2906 - 0.0015d_1 - 5.8369d_2 + 0.0006d_1d_2 \quad (6.4)$$

where

$$1500 \leq d_1 \leq 2500$$
$$1 \leq d_2 \leq 3$$

Sample Code of Roulette Wheel variation is given in the appendix. All other executable codes for all the chapters are given on the following website: https://sites.google.com/site/oatresearch.

6.3 Numerical Results and Discussion

Multi-CI, the variations of CI, and PSO algorithms were coded in MATLAB R2017 on Windows Platform with an Intel Core i3 processor and 4 GB RAM. Refer to the appendix section for the MATLAB codes of Multi-CI and variations of CI algorithm. For GA and SA algorithms, optimization toolbox is used from the same MATLAB version. The control parameters associated with the Multi-CI, variations of CI for solving the micro milling problem is listed in Table 6.1. Every problem is solved 30 times. The solutions obtained using variations of CI such as follow best, follow better, roulette wheel and alienation along with Multi-CI, GA, SA, and PSO have been shown in Table 6.2.

Similar to EDM and AWJM problems, micro milling problems are also inseparable, multi modal and nonlinear in nature. In Multi-CI exploration and exploitation mechanism is strong as every candidate search for the best solution from its own cohort as well as other cohorts. Due to this there are less chances of getting trapped into local minima which results in the faster convergence. For micro milling, with

Table 6.1 Control parameters and stopping criteria

Solution methodology	Parameter	Stopping criteria
Multi–CI	No. of cohorts = 3	Objective function value is less than 10^{-16}
	No. of candidates = 5	
	Value of reduction factor = 0.99	
	Behavioral variations for best candidates = 5 and for rest of the candidates = 10	
Variations of CI	No. of candidates = 5	
	Value of reduction factor = 0.99	
SA	Annealing function = Fast annealing	
	Re-annealing Interval = 100	
	Temperature update function = Exponential	
	Initial temperature = 100	
GA	Population size = 50	
	Scaling function = Rank	
	Selection = Stochastic uniform	
	Crossover fraction = 0.8	
	Mutation = Adaptive feasible	
	Crossover function = Heuristic	
PSO	Inertia co-efficient = Max 0.9 Min 0.2	
	Acceleration co-efficient = 2	

Table 6.2 Comparison of algorithms for solving R_a and M_t for micro milling

Micro Machining processes	Tool specification/cutter diameter	Objective function		Algorithms applied							Multi-CI
				GA	SA	PSO	Variations of CI				
							Roulette wheel	fbest	fbetter	Alienation	
Micro milling	0.7 mm	R_a	Mean	0.00	0.13	0.00	0.00	0.12	0.19	0.00	0.00
			S.D.	0.00	0.00	0.00	0.00	0.01	0.00	0.00	0.00
			Best	0.00	0.13	0.00	0.00	0.09	0.19	0.00	0.00
			Run time	1.40	2.60	1.12	0.06	0.05	0.06	0.14	0.21
		M_t	Mean	3.35	3.42	3.34	3.35	3.35	3.35	3.35	3.35
			S.D.	0.01	0.00	0.00	0.00	0.01	0.01	0.00	0.00
			Best	3.35	3.42	3.35	3.35	3.35	3.35	3.35	3.35
			Run time	1.44	2.62	0.74	0.04	0.04	0.04	0.10	0.16
	1 mm	R_a	Mean	0.03	0.16	0.03	0.03	0.11	0.21	0.03	0.03
			S.D.	0.00	0.00	0.00	0.01	0.02	0.00	0.01	0.00
			Best	0.03	0.15	0.03	0.03	0.06	0.21	0.03	0.03
			Run time	1.78	2.76	0.98	0.06	0.05	0.06	0.11	0.39
		M_t	Mean	3.23	3.47	3.23	3.23	3.23	3.23	3.23	3.23
			S.D.	0.00	0.02	0.00	0.00	0.00	0.00	0.00	0.00
			Best	3.23	3.44	3.23	3.23	3.23	3.23	3.23	3.23
			Run time	1.72	2.77	1.07	0.05	0.06	0.05	0.10	0.37

Mean Mean solution; *SD* Standard-deviation; *Best* Best solution; *Runtime* Mean runtime in seconds

0.7 mm and 1 mm tool, the performance of Multi-CI exceeded SA, follow best, and follow better variations of CI.

It is evident from Table 6.2 that the solutions to micro milling problem using GA are comparatively less robust with higher standard deviation. This is because, the population in GA is updated at every generation to simultaneously obtain the set of solutions. This iterative population generation along with the probabilistic rules to further guide the local search, the population converges to the approximate solution in the close neighborhood of the optimum solution. This generally is resulted in comparatively less robustness and higher standard deviation. In solving the micro-machining problems, every candidate in the Multi-CI algorithm competed with its own local best behavior as well as the best behavior chosen from the other cohorts. It resulted in faster convergence of every cohort solutions as compared to others. It is evident from the convergence plot in Fig. 6.3, follow better and follow best variation of CI converged to optimum solution with initial fluctuating pattern. This under-scored the efforts of the candidates in jumping out of the local minima. Similar to the performance of Multi-CI, the variations of CI algorithm exhibited more robustness as compared with other algorithms. In addition, Multi-CI, variations of CI, GA, SA, and PSO solution plots micro-milling with different tool diameters are exhibited in Fig. 6.3a–d respectively.

It is important to mention that for micro milling with 0.7 mm tool diameter 2% minimization of machining time was attained when compared to SA. Similarly, for 1 mm tool diameter, 6% minimization of machining time was achieved as compared to SA (refer to Table 6.2).

6.4 Summary

In this chapter, real world applications of variations of CI and Multi-CI algorithms have been successfully demonstrated in the advanced manufacturing domain by solving nonlinear, non-separable and multimodal problems. The algorithms of variations of CI and Multi-CI exhibited excellent performance against GA, SA, and PSO solving the micro milling problem. The robustness of the variations of CI as well as Multi-CI has been established for solving the real-world problems. The responses such as surface roughness and the machining time for micro milling process were minimized. It is important to mention that, two different milling cutter diameters viz. 0.7 mm, and 1 mm have been used. The results of variations of CI, and Multi-CI have been compared with contemporary algorithms such as GA, SA, and PSO in terms of solution quality, standard deviation, and run time. The results obtained using these methodologies are comparable with GA, and PSO. However, the algorithms outperformed SA. In near future, authors intend to apply CI and Multi-CI for more complex and constrained micro milling process problems. In the next chapter micro-drilling process is discussed. The process is performed on brass C360 workpiece with four pertinent drilling cutter diameters viz. 0.5, 0.6 0.8 and 1 mm for minimization of burr thickness and burr height.

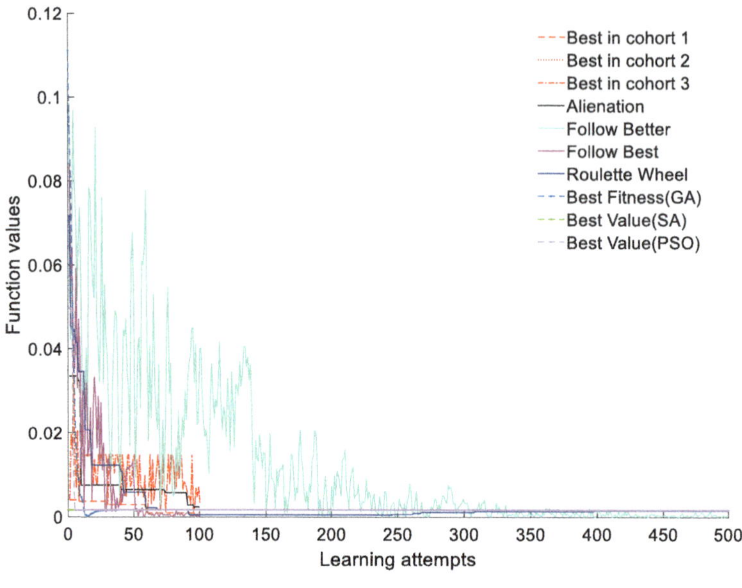

(a) Best solutions for Ra (Milling -Tool diameter 0.7mm)

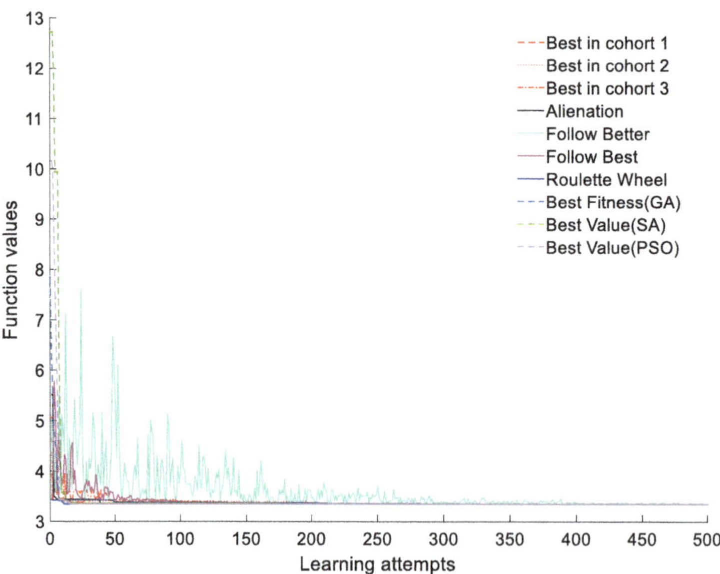

(b) Best solutions for M_t(Milling -Tool diameter $0.7mm$)

Fig. 6.3 Convergence plot for micro milling process

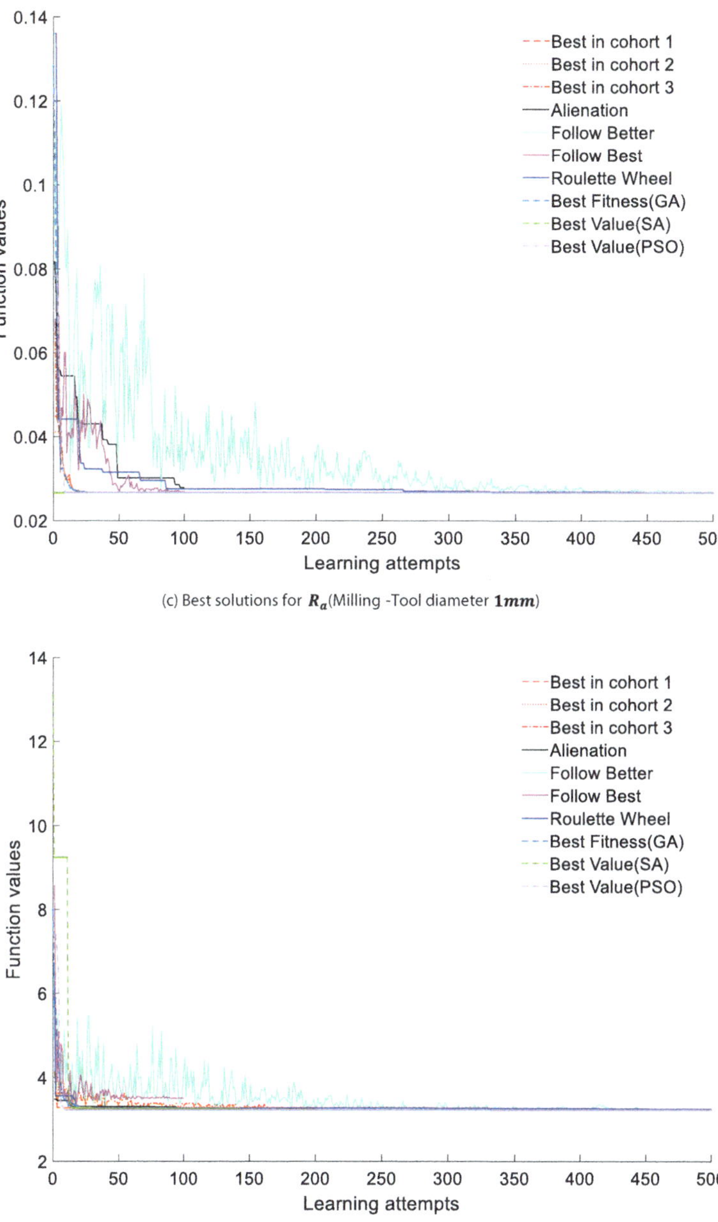

(c) Best solutions for R_a(Milling -Tool diameter $1mm$)

(d) Best solutions for M_t(Milling -Tool diameter $1mm$)

Fig. 6.3 (continued)

References

1. Bao W, Chen P, Tansel I, Reen NS, Yang S, Rincon D (2003) Selection of optimal cutting conditions by using the genetically optimized neural network system (GONNS). In: Kaynak O, Alpaydin E, Oja E, Xu L (eds) Artificial neural networks and neural information processing—ICANN/ICONIP 2003. ICANN 2003, ICONIP 2003. Lecture Notes in Computer Science, vol 2714. Springer, Berlin, Heidelberg
2. Gopalsamy BM, Mondal B, Ghosh S (2009) Taguchi method and ANOVA: An approach for process parameters optimization of hard machining while machining hardened steel
3. Hayajneh MT, Tahat MS, Bluhm J (2007) A study of the effects of machining parameters on the surface roughness in the end-milling process. Jordan J Mech Ind Eng 1(1):1–5
4. Jain VK (2008) Advanced (non-traditional) machining processes. In: Machining. Springer, London, pp 299–327
5. Kulkarni AJ, Durugkar IP, Kumar M (2013) Cohort intelligence: a self supervised learning behavior. In: 2013 IEEE international conference on systems, man, and cybernetics. IEEE, pp 1396–1400
6. Kumar SL, Jerald J, Kumanan S, Aniket N (2014) Process parameters optimization for micro end-milling operation for CAPP applications. Neural Comput Appl 25(7–8):1941–1950
7. Lu X, Jia Z, Wang H, Si L, Liu Y, Wu W (2016) Tool wear appearance and failure mechanism of coated carbide tools in micro-milling of Inconel 718 super alloy. Ind Lubric Tribol 68(2):267–277
8. Patankar NS, Kulkarni AJ (2018) Variations of cohort intelligence. Soft Comput 22(6):1731–1747
9. Pham DT, Elkaseera AM, Popova KP, Dimova SS, Olejnikc L, Rosochowskid A (2007) An experimental and statistical study of the factors affecting surface roughness in the micro milling process. In: Innovative Production machines and systems
10. Rahman M, Kumar AS, Prakash JRS (2001) Micro milling of pure copper. J Mater Process Technol 116(1):39–43
11. Saravanan M, Ramalingam D, Manikandan G, Kaarthikeyen RR (2012) Multi objective optimization of drilling parameters using genetic algorithm. Procedia Eng 38:197–207
12. Shastri AS, Kulkarni AJ (2018) Multi-cohort intelligence algorithm: an intra-and inter-group learning behaviour based socio-inspired optimisation methodology. Int J Parallel Emergent Distrib Syst 33(6):675–715
13. Shastri AS, Nargundkar A, Kulkarni AJ (2020) Multi-Cohort intelligence algorithm for solving advanced manufacturing process problems. Neural Comput Appl. https://doi.org/10.1007/s00521-020-04858-y
14. Zain AM, Haron H, Sharif S (2011) Estimation of the minimum machining performance in the abrasive water jet machining using integrated ANN-SA. Expert Syst Appl 38(7):8316–8326

Chapter 7
Optimization of Micro Drilling Process

Mechanical micro-drilling is one of the most widely used methods among several micro-hole making methods because of its least dependency on the material properties. Various factors such as tool diameter, spindle speed, tool helix angle, twist angle and feed rate determine the hole quality, and thus, they have to be chosen very carefully. Controlling burr formation in micro holes is significant as it causes deterioration of surface quality which reduces product durability and precision, assembly problems, wear and tear on the surface, etc. This chapter is based on the optimization of micro drilling process performed on brass C360 workpiece with four pertinent drilling cutter diameters viz. 0.5, 0.6, 0.8 and 0.9 mm for minimization of burr thickness and burr height. A regression model developed with the experimentation is adopted and socio inspired optimization algorithms variations of Cohort Intelligence (CI) and Multi-Cohort Intelligence (Multi-CI) have been applied [13]. Results achieved using variations of CI and Multi-CI algorithms have been presented and solutions are compared with Genetic Algorithm (GA), Simulated Annealing (SA), and Particle Swarm Optimization (PSO).

7.1 Introduction

The conventional machining processes have sufficed the requirements of the industries over the decades; however, the need for meso (500 μm–10 mm) and micro scale (1–500 μm) products are hastily increasing in various fields such as aeronautical, biomedical, automobile, optical and nuclear and semiconductor sector [3]. The associated materials are generally ceramics, metals, polymers, etc. Micro manufacturing is the key technology that can ensure the realization of miniature features in a product. Mechanical micromachining is further classified into conventional tool

© The Editor(s) (if applicable) and The Author(s), under exclusive license
to Springer Nature Singapore Pte Ltd. 2021
A. Shastri et al., *Socio-Inspired Optimization Methods for Advanced
Manufacturing Processes*, Springer Series in Advanced Manufacturing,
https://doi.org/10.1007/978-981-15-7797-0_7

based and advanced micromachining/nano finishing techniques. Conventional tool based micromachining techniques include micro turning, micro drilling and micro milling processes. There are several parameters associated with these processes which need to be controlled in order to achieve the desired responses.

Drilling is one of the basic machining processes of making holes and it is essentially for manufacturing industry like Aerospace industry, watch manufacturing industry, Automobile industry, medical industries and semiconductors. Especially Drilling is necessary in industries for assembly related to mechanical fasteners. It is reported that around 55,000 holes are drilled as a complete single unit production of the Air bus A350 aircraft. Micro drilling of metals is increasingly required as products become smaller and more highly functional. With increasing demand for precise micro component production, the importance of micro-hole drilling processes is increasing rapidly. Because of the requirement of deeper and smaller holes required in the above said industries, it is required for micro drilling process technologies to achieve higher accuracy and higher productivity [5].

Mechanical micro-drilling is one of the most widely used methods among several micro-hole making methods because of its least dependency on the material properties. Micro-drilling has a wide variety of applications associated with PCB circuits, microprocessors, in automotive industry for making micro-holes in fuel injectors, fasteners for micro-jacks and micro-pins. It is challenging to obtain good surface finish with reduced burr height in micro drilling. Various factors such as tool diameter, spindle speed, tool helix angle, twist angle and feed rate control the hole quality, and thus, they have to be chosen very carefully. Controlling burr formation in micro holes is significant as it causes deterioration of surface quality which reduces product durability and precision, assembly problems, wear and tear on the surface, etc. For macro-level drilling, burrs could be removed with de-burring; however, for micro-holes removing the burr becomes challenging because of poor accessibility of burr area and tight tolerance. In the past, numerous experimental and theoretical studies have been conducted for studying burr formation in drilling [4, 8, 14, 15]. Rahman et al. [11] investigated the effect of drilling parameter such as spindle speed, feed rate and drilling tool size on MRR, surface finish R_a, dimensional accuracy and burr thickness. Kilickap [6], employed Taguchi and RSM approaches for minimizing the burr height and the surface roughness in drilling of Al-7075. This work is an investigation of the influence of cutting parameters, such as cutting speed and feed rate, and point angle on burr height and surface finish produced. Zheng [16] used micro-drills with diameters of 0.1–0.3 mm to drill printed circuit boards with a highest spindle speed of 300,000 RPM. Aziz et al. [1] investigated the effects for minimization of burr formation and improvement of hole surface finish in micro drilling. Pansari et al. [9] experimented micro drilling of brass, and investigated and optimized the effect of cutting speed and feed rate on burr height and burr thickness. For comprehensive literature review, refer to the Chap. 1.

A socio inspired optimization technique Cohort intelligence (CI) was developed by Kulkarni et al. in [7]. This algorithm models the activity of the individuals of the society. Basically, it includes certain type of competition and interaction which eventually makes the society to evolve. Further, Patankar and Kulkarni [10] developed

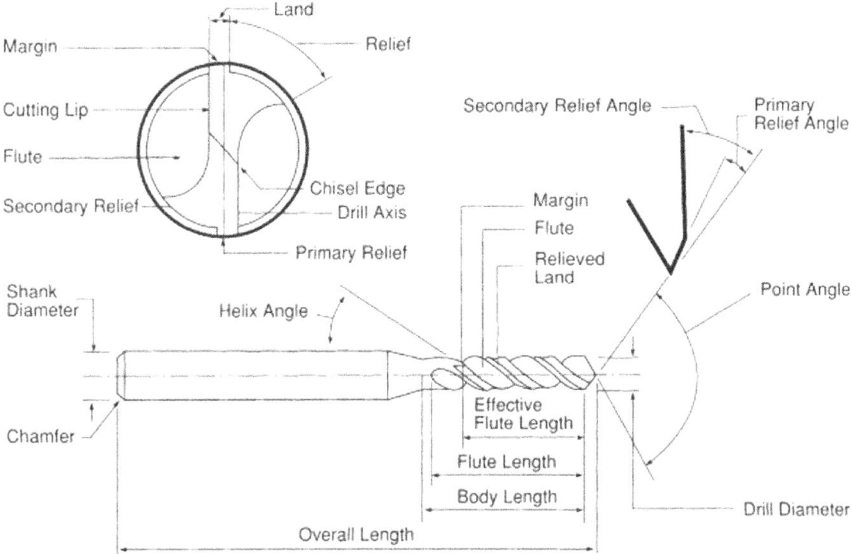

Fig. 7.1 Schematics of drilling cutter Choudhary [2]

variations of CI such as follow better (Fbetter), follow best (Fbest), follow worst, follow itself, follow median, alienation and roulette wheel (RW) (For details, refer Chap. 2). Multi-CI algorithm [12] is a modified version of CI in which intra and inter cohort learning mechanism amongst different candidates is modelled (For details, refer to the Chap. 3). The problem formulation for Micro Drilling process is given in the following section.

The schematics of drilling cutter is illustrated in Fig. 7.1. It shows the various geometries of a drill bit.

7.2 Problem Formulation

Pansari et al. [9] experimented micro drilling of brass and mathematically modelled the process. The same model has been used for optimization of micro drilling process. Burr height (B_h) and burr thickness (B_t) have been considered as performance responses with cutting speed (q_1 in rpm), feed (q_2 in $\frac{mm}{min}$) as process parameters. The developed regression model has four different drilling cutter diameters viz. 0.5, 0.6, 0.8 and 0.9 mm.

Equations (7.1)–(7.8) below shows the problem formulation:

For tool diameter 0.5 mm

$$Minimize\ B_h = 420.94 - 0.234q_1 - 99.91q_2 + 6.5510^{-5}q_1^2 + 22.152q_2^2 \quad (7.1)$$

$$Minimize\ B_t = 90.57 - 0.049q_1 - 27.12q_2 + 1.3210^{-5}q_1^2 + 5.54q_2^2 \tag{7.2}$$

For tool diameter 0.6 mm

$$Minimize\ B_h = 369.67 - 0.028q_1 - 156.79q_2 + 6.64 \times 10^{-6}q_1^2 + 23.162q_2^2 \tag{7.3}$$

$$Minimize\ B_t = 35.34 - 0.019q_1 - 0.59q_2 + 6.44 \times 10^{-6}q_1^2 + 0.51q_2^2 \tag{7.4}$$

For tool diameter 0.8 mm

$$Minimize B_h = 106.116 - 0.13q_1 - 6.62q_2 + 1.49 \times 10^{-6}q_1^2 + 4.75q_2^2 \tag{7.5}$$

$$Minimize B_t = 59.79 - 0.024q_1 - 11.3q_2 - 7.78 \times 10^{-6}q_1^2 + 2.18q_2^2 \tag{7.6}$$

For tool diameter 0.9 mm

$$Minimize\ B_h = 450.7 - 0.09q_1 - 38.48q_2 + 2.34 \times 10^{-5}q_1^2 + 5.03q_2^2 \tag{7.7}$$

$$Minimize\ B_t = 80.07 - 0.040q_1 - 14.81q_2 + 1.516 \times 10^{-5}q_1^2 + 4.65q_2^2 \tag{7.8}$$

where

$$1000 \leq q_1 \leq 2500$$
$$1 \leq q_2 \leq 4$$

Sample Code of Roulette Wheel variation is given in the appendix. All other executable codes for all the chapters are given on the following website: http://sites.google.com/site/oatresearch.

7.3 Numerical Results and Discussion

Multi-CI algorithm and the variations of CI were coded in MATLAB R2017 on Windows Platform with an Intel Core i3 processor and 4 GB RAM. Refer to the appendix section for the MATLAB codes of Multi-CI and variations of CI algorithm. For GA and SA algorithms, optimization toolbox is used from the same MATLAB version. The control parameters associated with the Multi-CI, variations of CI for solving the micro drilling problem is listed in Table 7.1. Every problem is solved 30 times. The solutions obtained using variations of CI such as follow best, follow better, roulette wheel and alienation along with Multi-CI, GA and SA have been shown in Table 7.2 (Fig. 7.2).

Table 7.1 Control parameters and stopping criteria

Solution methodology	Parameter	Stopping criteria
Multi-CI	No. of cohorts = 3	Objective function value is less than 10^{-16}
	No. of candidates = 5	
	Value of reduction factor = 0.99	
	Behavioral variations for best candidates = 5 and for rest of the candidates = 10	
Variations of CI	No. of candidates = 5	
	Value of reduction factor = 0.99	
SA	Annealing function = Fast annealing	
	Re-annealing interval = 100	
	Temperature update function = Exponential	
	Initial temperature = 100	
GA	Population size = 50	
	Scaling function = Rank	
	Selection = Stochastic uniform	
	Crossover fraction = 0.8	
	Mutation = Adaptive feasible	
	Crossover function = Heuristic	
PSO	Inertia co-efficient = Max 0.9 Min 0.2	
	Acceleration co-efficient = 2	

In Multi-CI exploration and exploitation mechanism is strong as every candidate search for the best solution from its own cohort as well as other cohorts. Due to this there are less chances of getting trapped into local minima which results in the faster convergence. Multi-CI outperformed SA in terms of solution quality as evident in Table 7.2. In addition, the convergence obtained by different variations of CI such as alienation, follow best, follow better and roulette wheel, Multi-CI, GA, SA, and PSO for objective function burr height and burr thickness for cutter diameters 0.5, 0.6, 0.8 and 0.9 mm respectively has been presented.

Different shapes of burrs were obtained after experiments for different speed and feed values. Broadly, there are three types of burrs, i.e., crown type, transient, and uniform burrs. Scanning Electron Microscope (SEM) images have been presented in Fig. 7.3. It has been observed that crown type burrs are formed at very low feed rate and high speed, while transient burrs are obtained at higher feed rate and higher speed. The uniform burrs are not solely dependent on feed rate and speed but tool diameter and tool type also, thus we calculated optimum values of feed rate and speed for different tools and results shows wide range of optimum values. For micro

Table 7.2 Comparison of Algorithms for solving B_h and B_t for micro drilling

Micro Machining processes	Tool specification/cutter diameter	Objective function		Algorithms applied							Multi-CI
				GA	SA	PSO	Variations of CI				
							Roulette wheel	Fbest	fbetter	Alienation	
Micro drilling	0.5 mm	B_h	Mean	99.29	134.13	99.29	99.29	99.29	99.29	99.29	99.29
			SD	0.00	1.30	0.00	0.00	0.00	0.00	0.00	0.00
			Best	99.29	131.79	99.29	99.29	99.29	99.29	99.29	99.29
			Run time	1.47	2.60	1.14	0.04	0.04	0.04	0.08	0.14
		B_t	Mean	11.91	21.13	11.90	11.91	11.91	11.91	11.91	11.91
			SD	0.00	1.22	0.00	0.00	0.00	0.00	0.00	0.00
			Best	11.91	18.14	11.90	11.91	11.91	11.91	11.91	11.91
			Run time	1.43	2.62	0.87	0.04	0.04	0.04	0.08	0.15
	0.6 mm	B_h	Mean	74.83	75.30	74.81	74.81	74.81	74.81	74.81	74.81
			SD	0.04	0.24	0.00	0.00	0.00	0.00	0.00	0.00
			Best	74.81	74.91	74.81	74.81	74.81	74.81	74.81	74.81
			Run time	1.62	2.65	0.76	0.04	0.04	0.04	0.08	0.14
		B_t	Mean	21.25	22.65	21.25	21.25	21.25	21.25	21.25	21.25
			SD	0.01	1.91	0.00	0.00	0.00	0.00	0.00	0.00
			Best	21.25	21.25	21.25	21.25	21.25	21.25	21.25	21.25

(continued)

Table 7.2 (continued)

Micro Machining processes	Tool specification/cutter diameter	Objective function		Algorithms applied							Multi-CI
				GA	SA	PSO	Variations of CI				
							Roulette wheel	Fbest	fbetter	Alienation	
			Run time	1.64	2.66	1.23	0.04	0.04	0.05	0.01	0.14
	0.8 mm	B_h	Mean	235.78	235.73	235.74	235.74	235.74	235.74	235.74	235.74
			SD	0.14	0.00	0.00	0.00	0.00	0.00	0.00	0.00
			Best	235.74	235.73	235.74	235.74	235.74	235.74	235.74	235.74
			Run time	1.73	2.67	1.34	0.06	0.05	0.06	0.12	0.17
		B_t	Mean	26.64	32.90	26.64	26.64	26.64	26.64	26.64	26.64
			SD	0.00	0.6	0.00	0.00	0.00	0.00	0.00	0.00
			Best	26.64	31.70	26.64	26.64	26.64	26.64	26.64	26.64
			Run time	1.68	2.71	0.88	0.035	0.036	0.04	0.076	0.16
	0.9 mm	B_h	Mean	305.07	307.74	305.07	305.07	305.07	305.07	305.07	305.07
			SD	0.00	2.46	0.00	0.00	0.00	0.00	0.00	0.00
			Best	305.07	305.39	305.07	305.07	305.07	305.07	305.07	305.07
			Run time	1.42	2.60	1.12	0.04	0.03	0.03	0.07	0.08

(continued)

Table 7.2 (continued)

Micro Machining processes	Tool specification/cutter diameter	Objective function		Algorithms applied			Variations of CI				Multi-CI
				GA	SA	PSO	Roulette wheel	Fbest	fbetter	Alienation	
		B_t	Mean	41.89	43.09	41.89	41.89	41.89	41.89	41.89	41.89
			SD	0.00	0.04	0.00	0.00	0.00	0.00	0.00	0.00
			Best	41.89	43.01	41.89	41.89	41.89	41.89	41.89	41.89
			Run time	1.47	2.64	0.96	0.04	0.03	0.04	0.07	0.10

Mean Mean solution; *SD* Standard-deviation of mean solution; *Best* Best solution

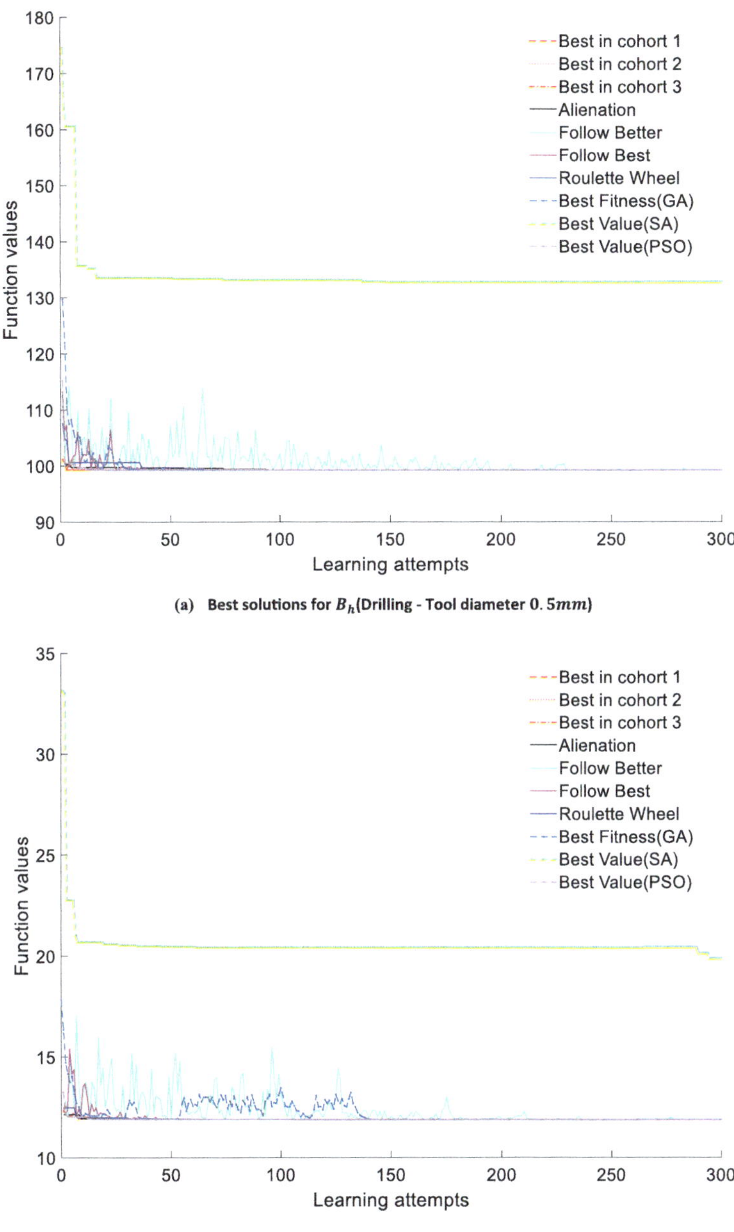

(a)   **Best solutions for B_h(Drilling - Tool diameter $0.5mm$)**

(b)   **Best solutions for B_t(Drilling -Tool diameter $0.5mm$)**

Fig. 7.2 Convergence plot for micro drilling process

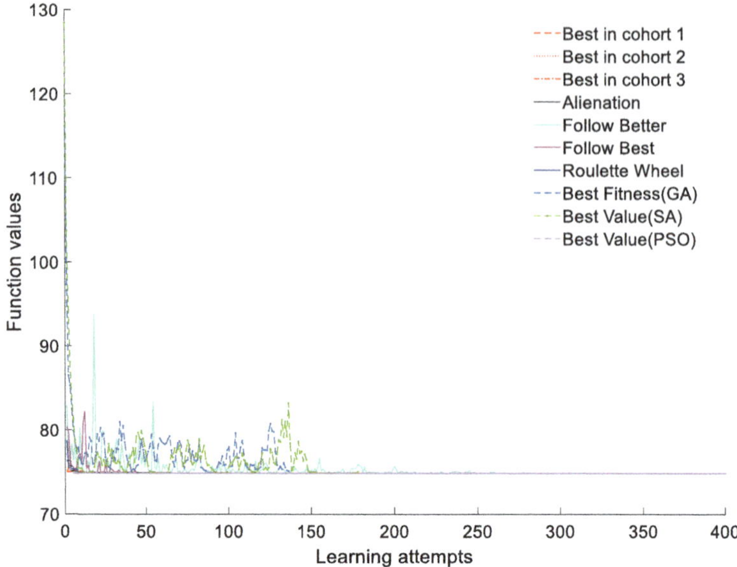

(c)Best solutions for B_h(Drilling - Tool diameter $0.6mm$)

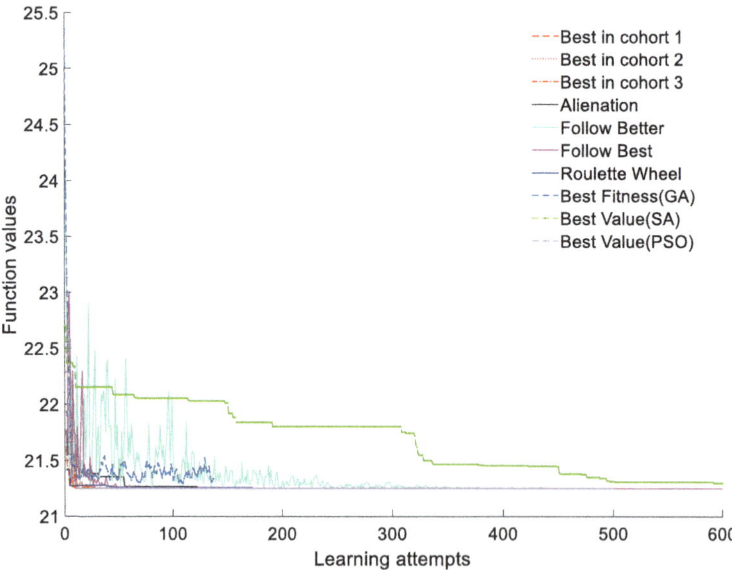

(d) Best solutions for B_t(Drilling -Tool diameter $0.6mm$)

Fig. 7.2 (continued)

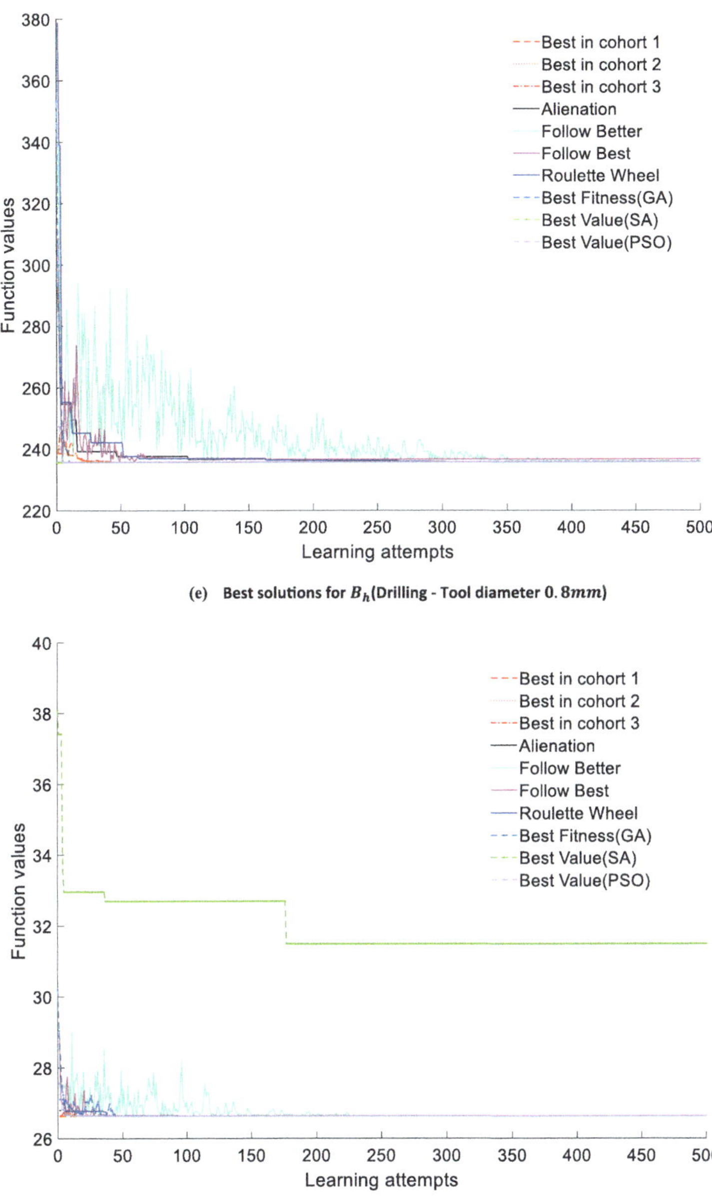

(e)   **Best solutions for B_h(Drilling - Tool diameter $0.8mm$)**

(f) **Best solutions for B_t(Drilling -Tool diameter $0.8mm$)**

Fig. 7.2 (continued)

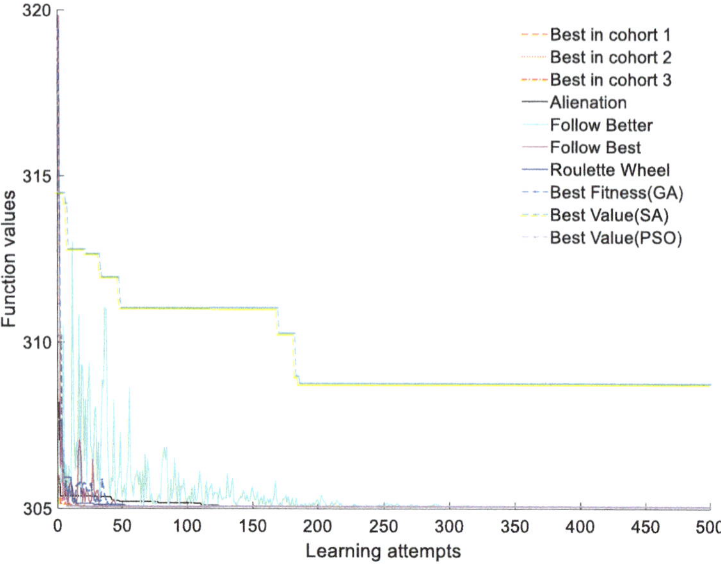

(g) Best solutions for B_h(Drilling - Tool diameter 0.9mm)

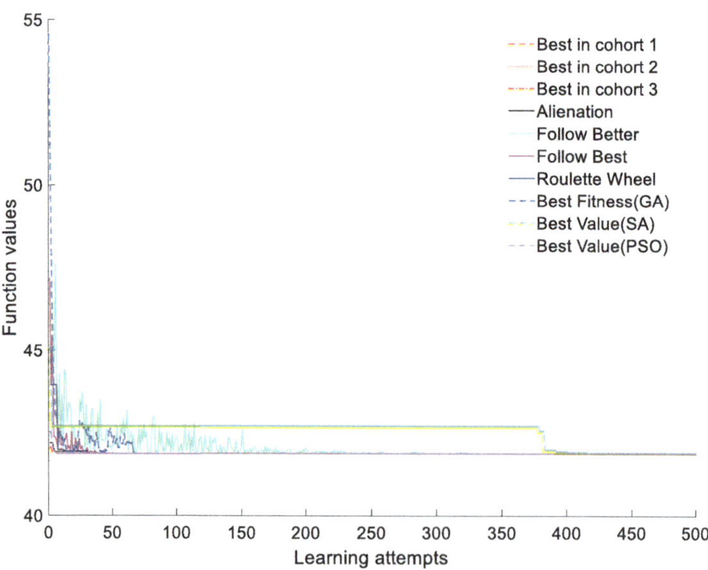

(h) Best solutions for B_t(Drilling -Tool diameter 0.9mm)

Fig. 7.2   (continued)

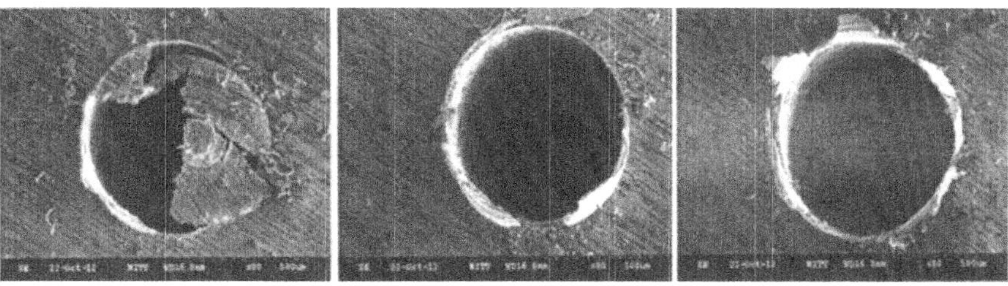

Fig. 7.3 SEM images of crown type burr, uniform type burr, transient type burr (left to right) [9]

drilling process, with tool diameter 0.5 mm, 24 and 34% minimization of B_h and B_t were achieved when compared to SA and GA. For tool diameter 0.8 and 0.9 mm, 16 and 3% minimization of B_t, respectively was achieved as compared to SA.

7.4 Summary

In this chapter, four variations of CI, Multi-CI, GA, SA, and PSO have been successfully applied on minimization of burr height and burr thickness for micro drilling process. It is important to mention that for micro drilling process, four different drilling cutter diameters viz. 0.5, 0.6, 0.8 and 0.9 mm had been used. The robustness of the variations of CI as well as Multi-CI has been established for solving the real world micro drilling optimization problems. With reference to the problem addressed, as four different cutter diameters are used, which gives a flexibility for choosing correct feed rate and spindle speed according to requirement, i.e., either minimum burr height or minimum burr thickness. For micro drilling process, Multi-CI yielded comparable results when compared with GA, PSO, and variations of CI. However, Multi-CI outperformed SA algorithm. It is observed during the experimentation and data analysis that in future, research can be done in the area of conventional micro-drilling at very high speeds as well as by varying tool twist angles and different tool materials. In the next chapter, cutting forces have been minimized for micro drilling of carbon fiber reinforced plastic (CFPR) composite materials for aerospace application.

References

1. Aziz M, Ohnishi O, Onikura H (2012) Innovative micro hole machining with minimum burr formation by the use of newly developed micro compound tool. J Manuf Process 14(3):224–232
2. Choudhary H (2007). Workshop technology. Media promoters and publishers. 10:608–620
3. Guo YB, Dornfeld DA (2000) Finite element modeling of burr formation process in drilling 304 stainless steel. J Manuf Sci Eng 122(4):612–619

4. Jain VK (2008) Advanced (non-traditional) machining processes. In: Machining. Springer, London, pp 299–327
5. Kirkpatrick S, Gelatt CD, Vecchi MP (1983) Optimization by simulated annealing. Science 220(4598):671–680
6. Kulkarni AJ, Durugkar IP, Kumar M (2013) Cohort intelligence: a self supervised learning behavior. In: 2013 IEEE international conference on systems, man, and cybernetics. IEEE, pp 1396–1400
7. Miyake T, Yamamoto A, Kitajima K, Tanaka Y, Takazawa K (1991) Study on mechanism of burr formation in drilling: deformation of material during burr formation. J Japan Soc Precision Eng 57(3):485–490
8. Özel T (2009) Editorial: special section on micromanufacturing processes and applications. Mater Manuf Process 24:12, 1235–1235. https://doi.org/10.1080/10426910903129349
9. Pansari S, Mathew A, Nargundkar A (2019) An investigation of Burr formation and cutting parameter optimization in micro-drilling of brass C-360 using image processing. In: Proceedings of the 2nd international conference on data engineering and communication technology. Springer, Singapore, pp 289–302
10. Patankar NS, Kulkarni AJ (2018) Variations of cohort intelligence. Soft Comput 22(6):1731–1747
11. Rahman AA, Mamat A, Wagiman A (2009) Effect of machining parameters on hole quality of micro drilling for brass. Modern Appl Sci 3(5):221–230
12. Shastri AS, Kulkarni AJ (2018) Multi-cohort intelligence algorithm: an intra-and inter-group learning behaviour based socio-inspired optimization methodology. Int J Parallel Emergent Distrib Syst 33(6):675–715
13. Shastri AS, Nargundkar A, Kulkarni AJ (2020) Multi-Cohort intelligence algorithm for solving advanced manufacturing process problems. Neural Comput Appl. https://doi.org/10.1007/s00521-020-04858-y
14. Shikata H, DeVries MF, Wu SM, Merchant ME (1980) An experimental investigation of sheet metal drilling. CIRP Ann 29(1):85–88
15. Takeyama H, Kato S, Ishiwata S, Takeji H (1993) Study on oscillatory drilling aiming at prevention of burr. J Japan Soc Precision Eng 59(10)
16. Zheng LJ, Wang CY, Fu LY, Yang LP, Qu YP, Song YX (2012) Wear mechanisms of micro-drills during dry high speed drilling of PCB. J Mater Process Technol 212(10):1989–1997

Chapter 8
Optimization of Micro Drilling of CFRP Composites for Aerospace Applications

In this chapter, variations of Cohort Intelligence (CI) algorithm have been applied for the minimization of cutting forces in x, y and z directions induced in micro drilling of carbon fiber reinforced plastic (CFPR) composite materials for aerospace applications. Three objective functions developed by Anand and Patra [4] have been used. These objective functions are cutting force, axial force, and thrust force induced in micro drilling of CFRP composites. Four variations of CI namely Follow Best, Roulette Wheel Follow Better, and Alienation have been applied. The algorithm was coded in MATLAB (R2016a). The solutions have been compared with the experimental work. The results obtained are much better than already available results giving significant reduction in cutting forces and thereby reducing power consumption and ultimately improvement in hole quality. As a future direction, other metaheuristics, socio-based algorithms could be applied for solving the problem. Also, variations of CI could be applied for solving constrained problems.

8.1 Introduction

The turning, milling and drilling processes are the fundamental subtractive machining processes. Micro-machining is the buzz word now a days as the whole world is moving towards the miniaturization. Micro machining finds the application in various domains. Micro drilling is widely applied process for manufacturing micro holes. It has been widely applied in electronics industry, in automotive industry, fabricating fastenings such as jacks and pins, etc. [14]. Factors like tooling parameters such as diameter, tool geometry, spindle speed, feed and work material govern the quality of holes produced, and thus they have to be chosen very carefully. When the dimensions are shifted from macro to micro, size effect comes into the consideration and hence

© The Editor(s) (if applicable) and The Author(s), under exclusive license
to Springer Nature Singapore Pte Ltd. 2021
A. Shastri et al., *Socio-Inspired Optimization Methods for Advanced
Manufacturing Processes*, Springer Series in Advanced Manufacturing,
https://doi.org/10.1007/978-981-15-7797-0_8

it becomes critical to achieve good process responses in micro machining processes. For micro drilling, good surface quality is vital for assembly purposes, however, achieving good surface inside the drilled hole is tough as the exposed surface is not enough to apply any polishing processes [10].

There are many process parameters such as cutting speed, tool geometry, cutting feed, etc. which governs the quality of hole and thus the selection and optimum combination of such parameters becomes critical. The burr formation phenomenon during micro drilling changes drastically as the internal surface is greatly affected by such burrs which might affect on product functionality and reliability [15]. For macro-level drilling, removing such burrs is relatively easy task as we can control the macro size hole. However, in case of micro, it is difficult because of the poor availability of area along with stringent tolerances applied. Hence, in the past various scientists have tried to study this process with experimental set up and theoretical studies [7].

Carbon fiber reinforced plastic (CFRP) composite material is an advanced composite which finds its practical application in various industry sectors such as aviation and automobile. Now a days steels and aluminium are getting replaced with the CFRP. It has properties such as high stiffness, high strength to weight ratio and good damping capacity. These practical products have typical micro hole features. The cutting mechanism in micro drilling is generally analyzed by studying traditional mechanical drilling with size effect. It is the downscaling of traditional mechanical drilling. During micro drilling process, the undeformed chip thickness reduces gradually, thus non linearly increasing the forces in radial direction. The add on effect of increased ploughing takes place at lower feed rates and thus tool wear increases or sudden tool breakage could take place. The process parameters such as feed, speed, ultrasonic vibrations produced in drills, and the workpiece fixturing were investigated by Wu et al. [16] for the delamination in CFRP drilling, focusing the internal delamination damages. Based on the investigation, the fixturing efficiency and the feed was investigated as well. The modeling of tooling wear with experimentation and theory analysis was provided by Zhang et al. [17]. A Finite Element Method has been applied to develop cutting force model. The model is based on the flank wear size and morphology. Later, experimentation was carried out for confirming the theoretical model, and the results show that proposed model can work well. Acoustic emission (AE) signals were applied to supervise AWJ machining by Pahuja and Ramulu [11]. The predictive approach was applied in simultaneous time-frequency Anand and Patra [4] performed experiments on CFRP composites with various speed and feed conditions to check the effect on cutting forces and hole quality. Response surface methodology was used for designing experiments. Analysis of variance was applied for identifying significant factor. Further, mathematical model was developed using regression analysis for the cutting forces.

Various nature inspired algorithms have been proposed recently and applied to optimize the process parameters of machining of CFRP composites. An experimental investigation of micro drilling of CFRP laminates using K20 carbide drill by varying the drilling parameters such as spindle speed and feed rate to determine optimum cutting conditions was performed and Genetic Algorithm (GA) was used to optimize the parameters [8]. Machining of CFRP composites with HSS tool was experimentally studied and Teaching Learning Based Optimization (TLBO) method has been applied for optimizing the process parameters. Solutions were compared with GA [2]. Jaya algorithm have been applied for optimizing the process parameters of CFRP machining process and solutions were compared with TLBO and GA algorithms [3]. An integrated multi-objective optimization methodology combining principal component analysis (PCA), fuzzy inference system (FIS), and Taguchi method has been applied for optimization of drilling of CFRP composites [1].

Kulkarni et al. [9] developed Cohort intelligence (CI) a socio-based optimization approach in which group of people interact with one another to obtain the solution which is globally acceptable. Here, roulette wheel approach is used by the people referred as candidates in the algorithm to decide whom to follow in order to achieve best solution. In this way, supervised learning is achieved by the candidates and using reduction factor to shrink the sampling space, candidates are moving towards global solution and not getting trapped into local minima. Algorithm was validated by solving few benchmark problems. The best possible solution is achieved if function value converges to optimal solution for significant number of learning attempts. In addition to this, other than roulette wheel approach, seven different variations for follow mechanism for candidates have been proposed by Patankar and Kulkarni in [12]. The so far applications of CI algorithm include shell and tube heat exchanger optimization problem by Dhavle et al. in [5], and constrained mechanical engineering problems [13]. The variations of CI were applied on AWJM process parameter optimization by Gulia and Nargundkar [6]. In this chapter four variation of CI viz. follow better, alienation, follow best and roulette wheel are applied to optimize cutting force, axial force and thrust force in x, y and z direction for micro drilling of CFRP composites as shown in Fig. 8.1. The problem formulation for cutting force minimization is given in the following section.

8.2 Cutting Force Minimization Problem Formulation

As mentioned earlier, the critical process parameters for the micro drilling process are cutting speed (V) and feed (f). Cutting forces induced in the micro drilling process, needs to be minimized as it affects the tool life, product quality, and machine tool structure. In micro drilling, cutting forces are generated along the 3 axes viz. x, y and z. The micro drilling experiments performed, and regression model developed Anand and Patra [4], has been used in the current work as objective function. The values of process parameters cutting speed and feed are chosen from within 15.7–39.2 and 1–5, respectively. Objective functions for F_x, F_y and F_z represents the cutting forces along x, y and z directions respectively and are given below in Eqs. (8.1), (8.2) and (8.3), respectively.

$$Minimize\ F_x = 0.38607 - 0.01498 \times V - 0.13220 \times f + 0.0012 \times f \times V$$

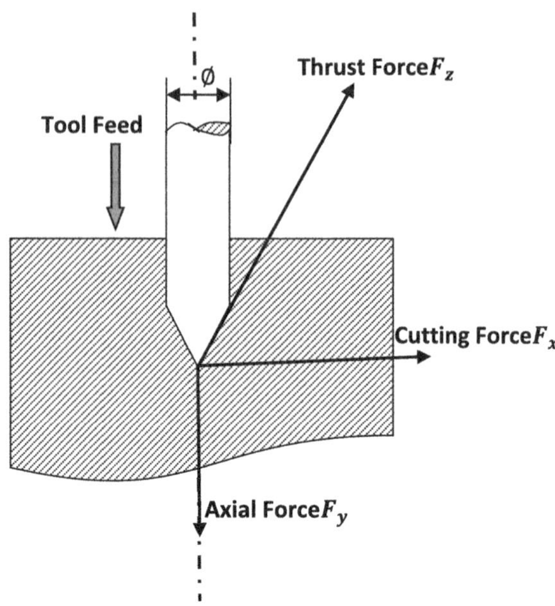

Fig. 8.1 Cutting forces in micro-drilling process

$$+\, 0.0003 \times V^2 + 0.019583 \times f^2 \tag{8.1}$$

$$Minimize\ F_y = 0.25371 - 0.00375 \times V - 0.12116 \times f + 0.00063 \times f \times V$$
$$+\, 0.000093 \times V^2 + 0.021 \times f^2 \tag{8.2}$$

$$Minimize\ F_z = 0.64841 + 0.078424 \times V + 0.59493 \times f - 0.0017 \times f \times V$$
$$-\, 0.0009 \times V^2 - 0.0204 \times f^2 \tag{8.3}$$

Sample Code of Roulette Wheel variation is given in the appendix. All other executable codes for all the chapters are given on the following website: sites.google.com/site/oatresearch.

8.3   Numerical Results and Discussion

The algorithm was coded in MATLAB R2016 on Windows Platform with an Intel Core i5 processor and 4 GB RAM. The control parameters associated with the variations of CI for solving the micro drilling of CFRP problem are listed in Table 8.1. To make the results statistically significant and robust, every problem is solved 30 times.

In Table 8.2, best and mean solutions for F_x, F_y and F_z of CFRP problem along

Table 8.1 Control parameters and stopping criteria

Solution methodology	Parameter	Stopping criteria
Variations of CI	No. of candidates = 5	Objective function value is less than 10^{-16}
	Value of Reduction factor = 0.99	

Table 8.2 Statistical solutions to problems using variations of CI

Machining Process	Objective function		Variations of CI			
			Roulette wheel	Follow best	Follow better	Alienation
Micro drilling	F_x	Mean	0.0569	0.0569	0.0569	0.0569
		SD	0.0000	0.0000	0.0000	0.0000
		Best	0.0569	0.0569	0.0569	0.0569
		Run ttme	0.73	0.94	0.78	0.79
	F_y	Mean	0.0704	0.0704	0.0704	0.0704
		SD	0.0000	0.0000	0.0000	0.0000
		Best	0.0704	0.0704	0.0704	0.0704
		Run time	0.98	0.63	0.96	0.83
	F_z	Mean	2.2057	2.2057	2.2057	2.2057
		SD	0.0000	0.0000	0.0000	0.0000
		Best	2.2057	2.2057	2.2057	2.2057
		Run time	0.81	0.76	0.64	0.82

Mean Mean solution, *SD* Standard-deviation of mean solution, *Best* Best solution, *Run Time* Mean runtime in seconds

with associated run time and standard deviation obtained using Roulette Wheel, Follow Best, Follow Better, and Alienation approach are shown. The solutions obtained using variations of CI have been compared with the existing experimental results in Table 8.3. Since the considered CFRP problem is non-linear, and inseparable in nature, exploration and exploitation mechanism of every variation should be efficient enough to find the optimum solution without getting trapped into local minima.

Figure 8.2(a)–(d) shows convergence obtained by different variations of CI such as alienation, follow best, follow better and roulette wheel respectively. In follow

Table 8.3 Comparison of solutions

Objective function	Experimental	Roulette wheel	Follow best	Follow better	Alienation
F_x	0.0950	0.0569	0.0569	0.0569	0.0569
F_y	0.1080	0.0704	0.0704	0.0704	0.0704
F_z	0.1080	2.2057	2.2057	2.2057	2.2057

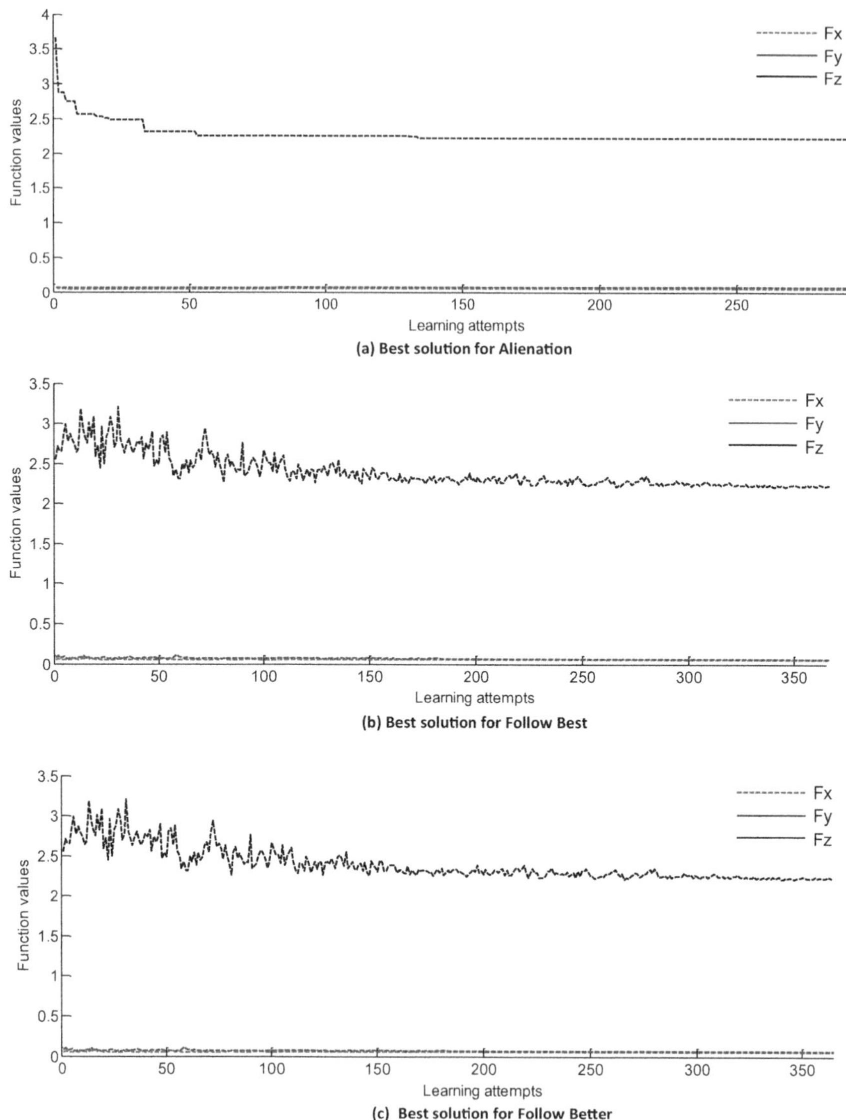

Fig. 8.2 Best solution plot

best and follow better approaches, candidates follow one of the candidates from the cohort. Hence, solutions are getting trapped in local minima as indicated in Fig. 8.2(b) and (c), as learning proceeds, the best and better candidate, jump out of local minima and eventually for all candidates the global minimum is obtained. Unlike follow better and follow best approaches, it is evident from Fig. 8.2(a) and (d), alienation and roulette wheel approach have shown significant difference in follow mechanism

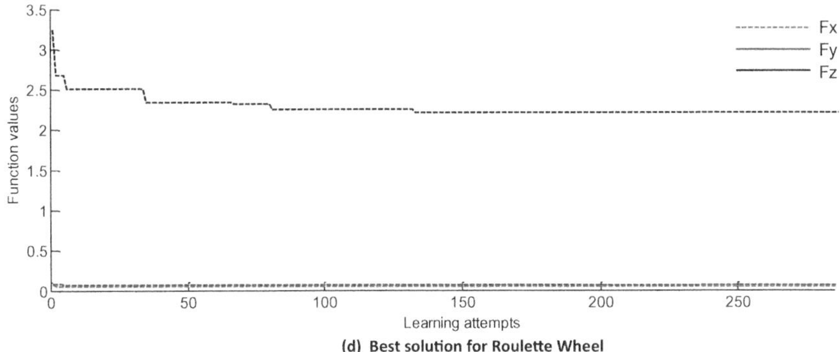

Fig. 8.2 (continued)

and results in global solution without trapping into local minima. It is evident from Table 8.3 that the results obtained using variations of CI are much better and robust as compared with experimental work.

8.4 Summary

In this chapter, four variations of socio inspired optimization algorithm CI have been successfully applied on cutting force reduction of micro drilling of CFRP composites for aerospace applications. Results have been compared with the experimental work results and have shown significant reduction in cutting forces in three directions. Due to reduction in cutting forces, tool life would be improved, tooling cost would be minimized, and high machining efficiency could be obtained. In the near future, authors intend to apply Multi-CI algorithm for solving the problem and also validate the simulation results using confirmation experiments.

References

1. Abhishek K, Datta S, Mahapatra SS (2015) Optimization of thrust, torque, entry, and exist delamination factor during drilling of CFRP composites. Int J Adv Manuf Technol 76(1–4):401–416
2. Abhishek K, Kumar VR, Datta S, Mahapatra SS (2017) Parametric appraisal and optimization in machining of CFRP composites by using TLBO (teaching–learning based optimization algorithm). J Intell Manuf 28(8):1769–1785
3. Abhishek K, Kumar VR, Datta S, Mahapatra SS (2017) Application of JAYA algorithm for the optimization of machining performance characteristics during the turning of CFRP (epoxy) composites: comparison with TLBO, GA, and ICA. Eng Comput 33(3):457–475
4. Anand RS, Patra K (2018) Cutting force and hole quality analysis in micro-drilling of CFRP. Mater Manuf Processes 33(12):1369–1377

5. Dhavle SV, Kulkarni AJ, Shastri A, Kale IR (2018) Design and economic optimization of shell-and-tube heat exchanger using cohort intelligence algorithm. Neural Comput Appl 30(1):111–125
6. Gulia V, Nargundkar A (2019) Optimization of process parameters of abrasive water jet machining using variations of Cohort Intelligence (CI). In: Applications of artificial intelligence techniques in engineering. Springer, Singapore, pp 467–474
7. Guo YB, Dornfeld DA (1999) Finite element modeling of burr formation process in drilling 304 stainless steel. J Manuf Sci Eng 122(4):612–619
8. Krishnaraj V, Prabukarthi A, Ramanathan A, Elanghovan N, Kumar MS, Zitoune R, Davim JP (2012) Optimization of machining parameters at high speed drilling of carbon fiber reinforced plastic (CFRP) laminates. Compos B Eng 43(4):1791–1799
9. Kulkarni AJ, Durugkar IP, Kumar M (2013) Cohort intelligence: a self supervised learning behavior. In: 2013 IEEE international conference on systems, man, and cybernetics. IEEE, pp 1396–1400
10. Miyake T, Yamamoto A, Kitajima K, Tanaka Y, Takazawa K (1991) Study on mechanism of burr formation in drilling: deformation of material during burr formation. J Japan So Precision Eng 57(3):485–490
11. Pahuja R, Ramulu M (2019) Surface quality monitoring in abrasive water jet machining of Ti6Al4V–CFRP stacks through wavelet packet analysis of acoustic emission signals. Int J Adv Manuf Technol 104(9–12):4091–4104
12. Patankar NS, Kulkarni AJ (2018) Variations of cohort intelligence. Soft Comput 22(6):1731–1747
13. Shastri AS, Thorat EV, Kulkarni AJ, Jadhav PS (2019) Optimization of constrained engineering design problems using cohort intelligence method. In: Proceedings of the 2nd international conference on data engineering and communication technology. Springer, Singapore, pp 1–11
14. Shikata H, DeVries MF, Wu SM, Merchant ME (1980) An experimental investigation of sheet metal drilling. CIRP Ann 29(1):85–88
15. Takeyama H, Kato S, Ishiwata S, Takeji H (1993) Study on oscillatory drilling aiming at prevention of burr. J Japan Soc Precision Eng 59(10)
16. Wu CQ, Gao GL, Li HN, Luo H (2019) Effects of machining conditions on the hole wall delamination in both conventional and ultrasonic-assisted CFRP drilling. Int J Adv Manuf Technol 104(5–8):2301–2315
17. Zhang Y, Wu D, Chen K (2019) A theoretical model for predicting the CFRP drilling-countersinking thrust force of stacks. Compos Struct 209:337–348

Chapter 9
Optimization of Micro-turning Process

The micro-turning processes have received a significant attention in the production of micro components with a diversity of materials including brass, aluminium, stainless steel, etc. Cutting speed, feed and depth of cut are the general process parameters/variables for micro turning process and surface roughness, flank wear, MRR, machining time are the typical process responses. This chapter is based on optimization of cutting speed, feed and depth of cut for the minimization of flank wear and surface roughness for micro turning process using 0.2 mm nose radius tool by applying variations of Cohort Intelligence (CI) and Multi-Cohort Intelligence (Multi-CI) algorithms [11]. The micro turning experiments and regression model developed by Durairaj and Gowri [1], has been adopted in the current work. The solutions of variations of CI, and Multi-CI have been compared with Genetic Algorithm (GA), Simulated Annealing (SA), and Particle Swarm Optimization (PSO) algorithms. The solutions exhibit comparable results.

9.1 Introduction

The conventional machining processes have sufficed the requirements of the industries over the decades; however, the need for meso (500 μm$-$10 mm) and micro scale (1$-$500 μm) products are hastily increasing in various fields such as aeronautical, biomedical, automobile, optical and nuclear and semiconductor sector [10]. The associated materials are generally ceramics, metals, polymers, etc. Micro manufacturing is the key technology that ensures the realization of miniature features in a product. Conventional tool based micromachining techniques include micro turning, micro drilling and micro milling processes. There are several parameters associated with these processes which need to be controlled in order to achieve the

© The Editor(s) (if applicable) and The Author(s), under exclusive license to Springer Nature Singapore Pte Ltd. 2021
A. Shastri et al., *Socio-Inspired Optimization Methods for Advanced Manufacturing Processes*, Springer Series in Advanced Manufacturing,
https://doi.org/10.1007/978-981-15-7797-0_9

desired responses. The micro-turning processes have received a significant attention in the production of micro components with a diversity of materials including brass, aluminium, stainless steel, etc. The working phenomenon for micro-turning is analogous to conventional machining operation. The material is removed from the workpiece by means of micro-tools which has specific characteristics due to its significant size reduction [6]. Durairaj and Gowri [1] experimented micro turning of Inconel alloy with cutting speed and feed as process parameters and developed regression model for optimizing surface finish and tool wear. Kumar [2] experimentally investigated micro turning of copper alloy with cutting speed and feed rate as process variables, surface finish and MRR being the responses. Sofuoğlu [7] optimized process parameters for unconventional turning process.

Micro turning process is governed by parameters/variables such as cutting speed, feed and depth of cut. The conventional model predicts that the surface roughness decreases with feed and fits well with the measured results, even in the micro turning process. But it is observed that surface roughness increases with feed decreases when the feed is in the range of micro scale. Tao et al. [12] has developed surface roughness prediction model based on peak to valley formation process. The prediction model has been verified with two groups of micro turning experiments. Results showed that size effect of specific cutting energy increases the surface roughness at small feeds. Piotrowska et al. [5] had developed mathematical model of micro turning process. With this model, it is possible for a machine operator, using the turning process parameters, to obtain a cutting model at very small depths of cut. Palani et al. [3] had developed on-line prediction of micro-turning multi-response variables by machine vision system using adaptive neuro-fuzzy inference system (ANFIS). Wu et al. [8] have presented an experimental investigation of specific cutting energy and surface quality based on the negative effective rake angle in micro turning process. A new model has been proposed to calculate the negative effective rake angle. The effective rake angle was found to be more negative with the decreasing ratio of uncut chip thickness to cutting edge radius. The minimum surface roughness was achieved near the critical negative rake angle. For comprehensive literature review, refer to the Chap. 1.

A socio inspired optimization technique Cohort intelligence (CI) was developed by Kulkarni et al. in [13]. This algorithm models the activity of the individuals of the society. Basically, it includes certain type of competition and interaction which eventually makes the society to evolve. Further, Patankar and Kulkarni [4] developed variations of CI such as follow better (Fbetter), follow best (Fbest), follow worst, follow itself, follow median, alienation and roulette wheel (RW) (For details, refer Chap. 2). Multi-CI algorithm [9] is a modified version of CI in which intra and inter cohort learning mechanism amongst different candidates is modelled (For details, refer to the Chap. 3). The problem formulation for micro turning process is given in the following section.

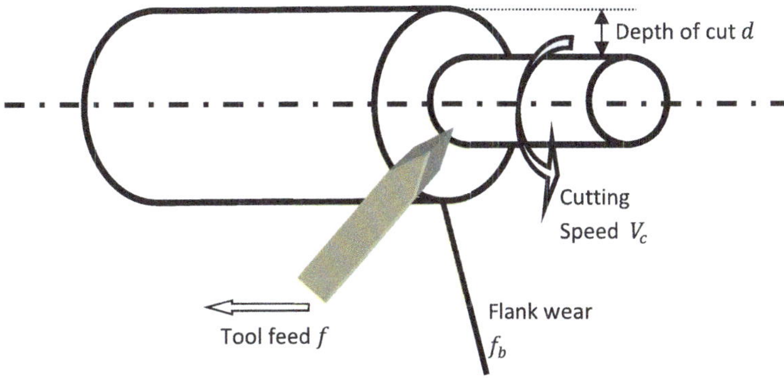

Fig. 9.1 Micro turning process

9.2 Problem Formulation

The micro turning experiments and regression model developed by Durairaj and Gowri [1], is adopted in the current work. Cutting speed (V_c in $\frac{m}{min}$), feed (f in $\frac{\mu}{rev}$) and depth of cut (d in μm) are considered as process parameters and flank wear (f_b) and surface roughness (R_a) are performance responses. The Fig. 9.1 describes the micro turning process with process parameters and responses.

The problem formulation is as follows:

$$Minimize \ f_b = 0.004 \times V_c^{0.495} \times f^{0.545} \times d^{0.763} \tag{9.1}$$

$$Minimize \ R_a = 0.048 \times V_c^{-0.062} \times f^{0.445} \times d^{0.516} \tag{9.2}$$

where

$$25 \leq V_c \leq 37$$
$$5 \leq f \leq 15$$
$$30 \leq d \leq 70$$

Sample Code of Roulette Wheel variation is given in the annexure. All other executable codes for all the chapters are given on the following website: sites.google.com/site/oatresearch.

9.3 Numerical Results and Discussion

Multi-CI algorithm [9], variations of CI [4], and PSO algorithm were coded in MATLAB R2017 on Windows Platform with an Intel Core i3 processor and 4 GB

RAM. Refer to the appendix section for the MATLAB codes of Multi-CI and variations of CI algorithm. For GA and SA algorithms, optimization toolbox is used from the same MATLAB version. The control parameters associated with the Multi-CI, variations of CI for solving the micro turning problem is listed in Table 9.1. Every problem is solved 30 times.

In Table 9.2, best and mean solutions for R_a, and F_b for micro turning problem along with associated run time and standard deviation obtained using variations of CI, Multi-CI, GA, SA, and PSO have been shown. Similar to other micro machining problems such as micro milling, and micro drilling, micro turning problem is non-linear, and inseparable in nature, exploration and exploitation mechanism of every variation should be efficient enough to find the optimum solution without getting trapped into local minima.

Figure 9.2a and b shows convergence obtained by different variations of CI such as alienation, follow best, follow better and roulette wheel, Multi-CI, GA, SA, and PSO for objective function R_a and F_b respectively. It is important to note that quick convergence of Multi-CI algorithm has been observed due to its intra and inter group

Table 9.1 Control parameters and stopping criteria

Solution methodology	Parameter	Stopping criteria
Multi-CI	No. of cohorts = 3	Objective function value is less than 10^{-16}
	No. of candidates = 5	
	Value of reduction factor = 0.99	
	Behavioral variations for best candidates = 5 and for rest of the candidates = 10	
Variations of CI	No. of candidates = 5	
	Value of reduction factor = 0.99	
SA	Annealing function =Fast annealing	
	Re-annealing interval = 100	
	Temperature update function = exponential	
	Initial temperature = 100	
GA	Population size = 50	
	Scaling function = Rank	
	Selection = Stochastic uniform	
	Crossover fraction = 0.8	
	Mutation = Adaptive feasible	
	Crossover Function = Heuristic	
PSO	Inertia coefficient = Max 0.9 Min 0.2	
	Acceleration coefficient = 2	

Table 9.2 Comparison of solutions for micro machining processes

Micro machining processes	Tool specification/cutter diameter	Objective function		Algorithms applied							Multi-CI
				GA	SA	PSO	Variations of CI				
							Roulette wheel	fbest	fbetter	Alienation	
Micro turning	0.2 mm nose radius	R_a	Mean	0.45	0.45	0.46	0.46	0.45	0.45	0.46	0.46
			SD	0.00	0.00	0.00	0.00	0.00	0.00	0.00	0.00
			Best	0.45	0.45	0.45	0.45	0.45	0.45	0.45	0.45
			run time	1.64	2.80	1.21	0.04	0.03	0.04	0.10	0.39
		f_b	Mean	0.63	0.64	0.63	0.63	0.63	0.63	0.63	0.63
			SD	0.00	0.00	0.00	0.00	0.00	0.00	0.00	0.00
			Best	0.63	0.64	0.63	0.63	0.63	0.63	0.63	0.63
			Run time	1.42	2.78	0.89	0.06	0.04	0.04	0.10	0.38

Mean Mean solution; *SD* Standard-deviation of mean solution; *Best* Best solution

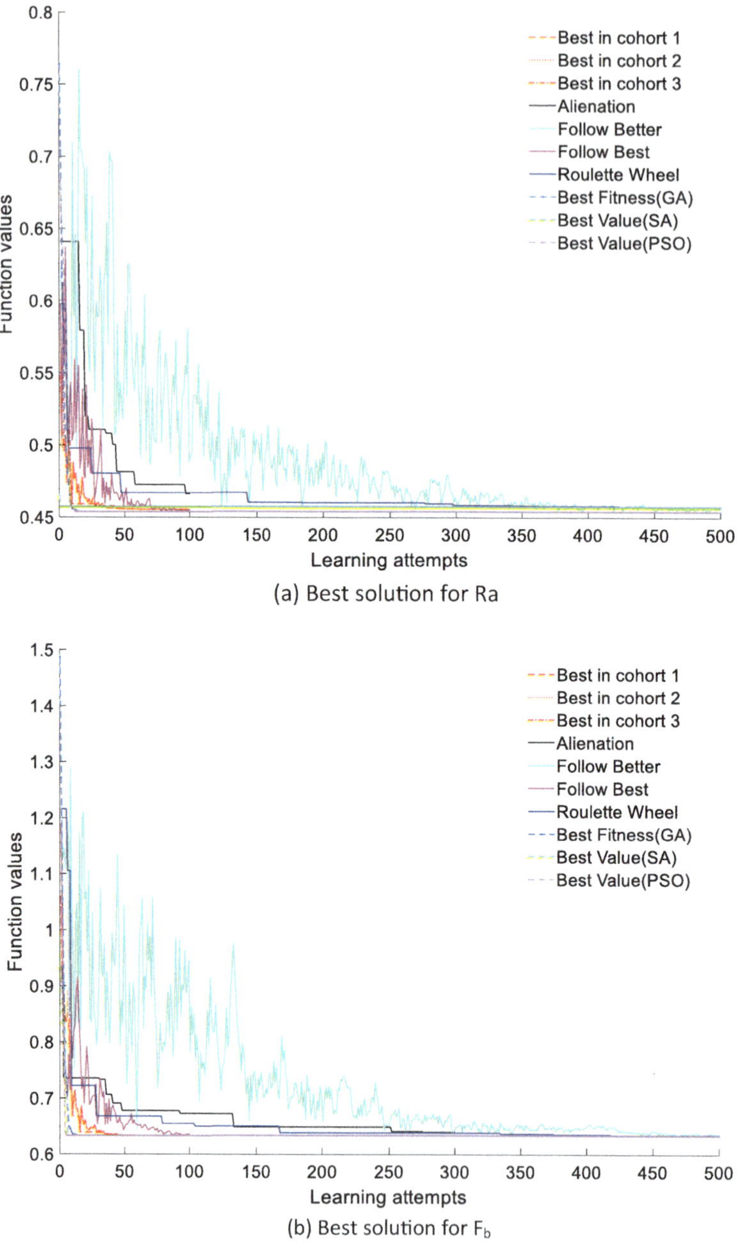

(a) Best solution for Ra

(b) Best solution for F_b

Fig. 9.2 Convergence plot for turning process

learning mechanism. In follow best and follow better approaches, candidates follow one of the candidates from the cohort. Hence, solutions are getting trapped in local minima as indicated in the Fig. 9.2, as learning proceeds, the best and better candidate, jump out of local minima and eventually for all candidates the global minimum is obtained. Unlike follow better and follow best approaches, alienation and roulette wheel approach have shown significant difference in follow mechanism and results in global solution without trapping into local minima. It is evident from Table 9.2 that results obtained using variations of CI and Multi-CI are comparable with GA, SA, and PSO.

9.4 Summary

In this chapter, the applicability of the socio-inspired optimization method referred to as Variations of CI, and Multi-Cohort Intelligence (Multi-CI) is validated by solving real world micro turning problem from advanced manufacturing domain. The minimization of surface roughness and flank wear for micro turning process with the tool having nose radius of 0.2 mm has been achieved. Multi-CI has shown comparable and robust results against GA, SA, PSO, and variations of CI. The comparative convergence of all the algorithms has been independently illustrated for both the objective functions. In the near future efforts could be made to develop constrained handling methods for CI, and Multi-CI algorithms to solve soft and hard constrained problems associated with a variety of domains including supply-chain management, transportation, dynamic control problems.

References

1. Durairaj M, Gowri S (2013) Parametric optimization for improved tool life and surface finish in micro turning using genetic algorithm. Procedia Eng 64:878–887
2. Kumar SL (2019) Measurement and uncertainty analysis of surface roughness and material removal rate in micro turning operation and process parameters optimization. Measurement 140:538–547
3. Palani S, Natarajan U, Chellamalai M (2013) On-line prediction of micro-turning multi-response variables by machine vision system using adaptive neuro-fuzzy inference system (ANFIS). Mach Vis Appl 24(1):19–32
4. Patankar NS, Kulkarni AJ (2018) Variations of cohort intelligence. Soft Comput 22(6):1731–1747
5. Piotrowska I, Brandt C, Karimi HR, Maass P (2009) Mathematical model of micro turning process. Int J Adv Manuf Technol 45(1–2):33–40
6. Robinson GM, Jackson MJ (2005) A review of micro and nanomachining from a materials perspective. J Mater Process Technol 167(2–3):316–337
7. Sofuoğlu MA, Çakır FH, Kuşhan MC, Orak S (2019) Optimization of different non-traditional turning processes using soft computing methods. Soft Comput 23(13):5213–5231

8. Wu X, Li L, Zhao M, He N (2016) Experimental investigation of specific cutting energy and surface quality based on negative effective rake angle in micro turning. Int J Adv Manuf Technol 82(9–12):1941–1947
9. Shastri AS, Kulkarni AJ (2018) Multi-cohort intelligence algorithm: an intra-and inter-group learning behaviour based socio-inspired optimisation methodology. Int J Parallel Emergent Distrib Syst 33(6):675–715
10. Özel T (2009) editorial: special section on micromanufacturing processes and applications, Mater Manuf Processes 24(12):1235–1235. https://doi.org/10.1080/10426910903129349
11. Shastri AS, Nargundkar A, Kulkarni AJ (2020) Multi-Cohort intelligence algorithm for solving advanced manufacturing process problems. Neural Comput Appl. https://doi.org/10.1007/s00 521-020-04858-y
12. Tao W, Zhong Y, Feng H, Wang Y (2013) Model for wear prediction of roller linear guides. Wear 305(1–2):260–266
13. Kulkarni AJ, Durugkar IP, Kumar M (2013) Cohort intelligence: a self supervised learning behavior. In: 2013 IEEE international conference on systems, man, and cybernetics. IEEE, pp 1396–1400

Appendix

Cohort Intelligence Code with Roulette Wheel Selection variation for Surface Roughness (R_a) and *kerf* of Abrasive Water Jet Machining (AWJM) is given below. Other executable codes for all the chapters are given on the following website: https://sites.google.com/site/oatresearch.

```matlab
clear all
clc
tic
c = 5;   % number of candidates
n = 4;   % number of dimensions
upper_bound = [1.25 1.5 96 600];
lower_bound = [0.9 0.95 20 200];
r = 0.99;
max_iterations = 500;
for cnt_runs = 1:1
x = (upper_bound - lower_bound).*rand(c,n) + lower_bound; % random value generation
range = upper_bound - lower_bound;                       % original range calculation
cnt_iterations = 1;

    while (cnt_iterations <= max_iterations) % loop for max iterations

        for cnt_candi = 1:c
           Fun(cnt_candi,:)= Ra_awjm(x(cnt_candi,1:n)); % call objective function Ra, Kerf here
        end

        inverse_F = Fun.^-1;                  % 1/fxn
        denominator_pc = sum(inverse_F,1);    %denominator calucated((1/fx(c1)+...+1/fx(cn)) for probability calculation

        for cnt_pc = 1:c
           prob(cnt_pc,1) = inverse_F(cnt_pc,1)/denominator_pc; % probability pc1,...,pcm
        end

        pcsum = sum(prob);  % checking cummulative sum = 1?

          for cnt_candi = 1:c
           c_follow(cnt_candi) = RouletteWheelSelection(prob);
          end

        [M,I] = min(Fun);
           Fbest(1,:) = M(1, :);

    xvect_best(:,:) = x(I(1),:);                    % x vector for all best

new_matrix = x(c_follow,:); % new matrix generation

range = r*range;

for cnt_column = 1:n
    if abs(range(cnt_column))> 1e-16
       new_range =   range/2;
    else
       new_range =   1e-16;
    end
end
```

```matlab
       for cnt_row = 1:c
         for cnt_column = 1:n
             new_lower_bound(cnt_row,cnt_column) = new_matrix(cnt_row,cnt_column)- new_range(cnt_column);
             new_upper_bound(cnt_row,cnt_column) = new_matrix(cnt_row,cnt_column)+ new_range(cnt_column);
         end
       end
        for cnt_row = 1:c
         for cnt_column = 1:n
             if new_lower_bound(cnt_row,cnt_column) < lower_bound(1,cnt_column)
                 new_lower_bound(cnt_row,cnt_column) = lower_bound(1,cnt_column);
             end
             if new_upper_bound(cnt_row,cnt_column) > upper_bound(1,cnt_column)
                 new_upper_bound(cnt_row,cnt_column) = upper_bound(1,cnt_column) ;
             end
         end
        end


     x = (new_upper_bound -new_lower_bound).*rand(c,n) + new_lower_bound;
     x(c,:) = xvect_best;



         if (cnt_iterations <= max_iterations)
         function_all(:,cnt_iterations) = abs(Fbest);
         cnt_all (cnt_iterations) = cnt_iterations;
         x_all(:,:,cnt_iterations) =  x;
         end
         hold on;
         cnt_iterations = cnt_iterations + 1;
    end
%%%%%%%%%%%%% Convergence Plot %%%%%%%%%%%%%%%%%%%%



%     if (rem(cnt_iterations,1)==0)
%         stop_1=1;
%     end
%     if (rem(cnt_iterations,1)==0)
%         stop_1=1;
%     end
%     figure(1)
%             xlabel('Learning attempts','FontSize',12)
%             ylabel('Behavior','FontSize',12)
%         plot(cnt_all,function_all(1,:),'k*');
%         plot(cnt_all,function_all(2,:),'ko');
%         plot(cnt_all,function_all(3,:),'k+');
%         plot(cnt_all,function_all(4,:),'k.');
%         plot(cnt_all,function_all(5,:),'ks');
%             legend ('candi1','candi2','candi3','candi4','candi5');
% %            %pause(2);
% %             hold on;
%
%
% %     end


toc;

time_elapsed = toc;

end
%%%%%%%%%%%%%%%%%%%%%%%%%%%%%% Ra  %%%%%%%%%%%%%%%%%%%%%%%%%%%%%%%%%%%

function y = Ra_awjm(x)
y1= -23.309555 + 16.6968*x(1)+ 26.9296*x(2)+0.0587*x(3)+0.0146*x(4);
y2=-5.1863*x(2)*x(2) -10.4571*x(1)*x(2) - 0.0534*x(1)*x(3)- 0.0103*x(1)*x(4)+ 0.0113*x(2)*x(3)- 0.0039*x(2)*x(4);
y = y1+y2;


%%%%%%%%%%%%%%%%%%%%%%%%%%%%%% kerf %%%%%%%%%%%%%%%%%%%%%%%%%%%%%%%%%%%

function y = kerf_awjm(x)
y1= -1.15146 + 0.70118*x(1)+ 2.72749*x(2)+ 0.00689*x(3)-0.00025*x(4)+0.00386*x(2)*x(3);
y2=-0.93947*x(2)*x(2)-0.25711*x(1)*x(2)-0.00314*x(1)*x(3)-0.00249*x(1)*x(4)+0.00196*x(2)*x(4)-0.00002*x(3)*x(4);
y = y1+y2;
end
%%%%%%%%%%%%%%%%%%%%%%%%%%%%%%%%%%%%%%%%%%%%%%%%%%%%%%%%%%%%%%%%%%%%%%%%
```

The manufacturer's authorised representative in the EU is Springer
Nature Customer Service Centre GmbH, Europaplatz 3, 69115 Heidelberg,
Germany. If you have any concerns regarding our products, please
contact ProductSafety@springernature.com

Printed and bound by CPI Group (UK) Ltd, Croydon, CR0 4YY
01/12/2025
02008619-0004